Values
and
Technology

Editor

Gabriel R. Ricci

Values
and
Technology

Religion & Public Life

Volume
37

Routledge
Taylor & Francis Group

LONDON AND NEW YORK

First published 2011 by Transaction Publishers

Published 2017 by Routledge
2 Park Square, Milton Park, Abingdon, Oxon OX14 4RN
711 Third Avenue, New York, NY 10017, USA

Routledge is an imprint of the Taylor & Francis Group, an informa business

ISSN: 1083-2270
ISBN 13: 978-1-4128-1118-7 (pbk)

Volumes 1 through 28 were originally published under the title *This World: An Annual of Religion and Public Life.*

Contents

Introduction

Gabriel R. Ricci

In 1749, Jean-Jacques Rousseau enthusiastically responded to an essay competition which asked authors to consider whether or not the arts and sciences had advanced human progress. His winning entry, *Discourse on the Arts and Sciences*, surprised leading Enlightenment thinkers who enthusiastically upheld the positive benefits of humanity's technological advance. Rousseau would go on to have uneasy relationships with his enlightenment cohort, though this did not prevent him from providing essays on music and political economy to Diderot's massive *Encyclopédie*. Voltaire, who celebrated and relished the accoutrements of civilization, mocked Rousseau's praise for an original unencumbered state of nature in which man enjoyed an optimum level of freedom. Voltaire was certain that Rousseau wished to restore a bestial and primitive state upon humanity, but Rousseau could not have been that naïve. Like Jefferson and ancient authors like Plutarch, Livy, and Virgil who laid the groundwork, Rousseau linked the virtuous life with the independent farmer, unencumbered by the constraints of civilization, whose innate skills were honed by self-reliance and an intimate understanding of the natural world. Rousseau's essay warned that the progressive intrusion of technology would so focus humanity's attention outward into an artificially produced social world that it would nullify original inclinations to self-examination and self-improvement. Over the centuries, Rousseau would be criticized for providing the conceptual germ of totalitarian sovereignty in his idea of the general will, being responsible for the excesses of the French Revolution as well as promoting the virtues of nationalism. Still Voltaire's mocking response to Rousseau's work remains the most biting. Rousseau would fire back at Voltaire with his critique of the latter's poem in response to the 1755 Lisbon earthquake. In keeping with his faith in the noble savage, Rousseau claimed that the devastation was socially produced and might have been mitigated by a more natural lifestyle unencumbered by possessions which, he argued, undermined evacuation efforts. Today their intellectual rivalry is muted by their mutual aggrandizement in France's Pantheon; Voltaire arrived in 1791, three years before Rousseau took occupancy.

The argument that engaged Voltaire and Rousseau has not been muted. Given the unprecedented intrusion of technology into our lives, the question raised by Rousseau's critique remains acute. The technological augmentation of the social structure has introduced new challenges to conventional morality and in some cases threatens to make some traditional ethical questions moot. Sigmund Freud, who like Rousseau questioned the effects of civilization on our happiness, noted that scientific innovations provided sufficient add-ons to make man a prosthetic God. Freud, like Enlightenment critics Max Horkheimer and Theodor Adorno, also acknowledged the connection between high levels of civilization and the exploitation and control of nature. As post-Second World War social critics, Horkheimer and Adorno interpreted the Holocaust as the culmination of the Enlightenment commodification of knowledge with its concomitant exploitation of nature and humanity. These subsequent critiques demand that we routinely examine the influence of science and technology whose frontiers have been stretched to nano-borders where control at atomic and molecular levels now attracts our technological and moral attention.

Like Horkheimer and Adorno, Martin Heidegger also drew our attention to the threatening powers of technology, particularly how the transformation of humanity and nature into what he dubbed a standing-reserve reduced the world to a collection of instrumental entities, thus concealing the nature of his perennial quest, Being itself. In a less cryptic style, Heidegger's estranged student, Hans Jonas, would later provide a framework for an ethics that superseded antiquated moral thinking which presumed a fixed human condition and a fixed natural world. The dominion of technology in the modern world and the invasive powers of human action, Jonas argued, demands a reconsideration of established conventions predicated on fixed ideas of nature and humanity. Jonas' prescription that we ought to act so that the effects of our actions are compatible with the permanence of authentic human life is embedded in Kantian ethics and echoes Aldo Leopold's challenge to act in such a way as to preserve the integrity of the biotic community. Such comprehensive exhortations challenge the moral imagination but their intended scope suggests a need for a radical re-evaluation of the consequences of our actions and the shortsightedness of our institutional goals and imperatives.

In a recently minted normative realm that Heidegger could not have anticipated, the work of Luciano Floridi has begun to address the need to frame an ethical discourse that can adequately address the problems attached to information technology. The first article of this collection titled *Values and Technology*, "Information Ethics: A Critical Assessment," by John Barker, delves into Floridi's judgment that in the context of information technology, there is no convenient way to draw the line between what ought to receive moral consideration and what lies outside our normative compass. If environmentalists, following Leopold, have stretched the conventional view of community, Floridi's work goes beyond the frontiers mapped out by ecologists and animal liberationists alike. Barker carefully scrutinizes Floridi's consideration of the inherent worth of what he calls "informational objects" since Floridi's conclusion, in effect, entails all of existence. What is at stake in Floridi's "informational ethic" is the preservation of Being itself. While we ordinarily conceive of information as instrumental and closer to the dynamic of the Kantian hypothetical imperative, Floridi, Barker shows, only takes impartiality to its logical conclusion. Consequently, Floridi recommends that the destruction and depletion of informational objects and any impoverishment of *being* is to be avoided. The details of Floridi's innovative ethics are clarified in Barker's fine introduction to his work and though we can recognize the Heideggerian backdrop to his system, the fact that information, on his terms, is everywhere demands the moral sensitivity and cultivation prescribed in Buddhist mindfulness.

Floridi's Information Ethic entails a metaphysics and may seem remote when we consider the ethical dimensions of a particular discipline or field. To what extent do lawyers address metaphysics in the legal realm? To what extent do business departments incorporate a metaphysical outlook when pressing out the ethical consequences of operating in a business environment? Even those who engage their students with the ethical aspects of cyberspace need not wait for Floridi et al. to map out the new terrain in information ethics. Fani Zlatarova provides a practical guide for instructors who introduce students to the ethical components of computer technology. Zlatarova's "Teaching Values in Computing Courses through Theory and Practice" begins to provide guidelines for an applied ethics curriculum for computing majors. She outlines a rigorous program incorporating the many facets of computing technology to advocate built-in ethical components to an activity that increasingly insinuates itself into our lives and even promises to make inroads into our physical selves. Zlatarova's efforts make us aware that the sensation of privacy behind the artificial glow of our screens is a window onto an integrated community that cannot ignore the principles of right and wrong.

Alan Kim's essay "Problems of Technology" begins with an argument which clarifies our fundamental relationship with technology or how it is that technology is an inherent facet of the human condition. Sometimes perceived as social construction, technology is construed as an appendage to humanity. Kim brings to bear the scientific evidence which suggests that our species should be more accurately labeled *homo faber.* The problems that technology presents, then, translates into problems of application and alienation. For Kim this is the essential problem of technology, which is resolved by overcoming the pitfalls of *imbalance.* The practical solution is provided in the example of a musician who is one with his instrument. Kim's conclusion belies his theoretical dependence on Aristotle's craft analogy which is grounded in a praxis which produces a mental predisposition that guarantees the unity of man and instrument, the integrity of *technē* and *aretē*, the essential ingredients of the ancient Greek moral formula.

William Cornwell likewise turns to the ancient world in his consideration of how pharmaceutical and technological interventions can positively enhance our cognitive powers. Anticipating various forms of cognitive augmentation that are staring us in the face, Cornwell speculates on how Aristotelian virtue ethics can be mechanically perfected. Cornwell is not sanguine about transhumanism, that we are a species in transition to a more powerful intelligence; he explores how responsible cognitive augmentation might help us develop greater moral and intellectual virtues. These thought experiments are familiar to us from the science fiction literature that explores dystopic futures in which better living through chemicals is institutionally sanctioned. With the therapeutic use of artificial retinas, Cornwell maintains, the mind/machine barrier is breaking down; augmenting intelligence is already underway in computer engineering and neuroscience. How such engineering can improve our moral status is indeed a provocative thesis. Cornwell assumes the burden of showing how an approach to virtue ethics, heavily dependent on practice and the cultivation of habitual behaviors and dispositions, can be artificially produced. Aristotle might balk, since his emphasis on praxis would become moot. However, it might very well be Aristotle's exploitation of the critical indices of pleasure and pain to produce and explain virtuous behavior, which will underlie the possibility that technology will be capable of intervening to adjust and regulate our behavior. But the anticipation of a severe punishment, including execution, has never proved a reliable deterrent. But, after Ray Kurzweil's prediction that computers will soon be able to pass the Turing test and provide more enhanced desirable sexual encounters, Cornwell's thesis is safely within philosophical bounds.

Grant Havers' essay "Natural Rights, Religion and the Biogenetic Revolution" explores the controversies surrounding the biogenetic explosion through an examination of the competing arguments which occupy the philosophical debate. Though he acknowledges the dominance of the principle of utility and the appeal of Kantian respect for persons, he is certain that the tradition of natural rights, which derives from considerations of the human essence, trumps all other perspectives. In particular, natural rights gains the higher moral ground by avoiding the economic calculation of life. Havers analyses the various traditions of natural rights: enlightenment liberalism, ancient Greek philosophy and the religious tradition. While we customarily associate natural rights with enlightenment political ideals and the way in which the American Founding incorporates those ideals, Havers finds more potent strains in Plato, Aristotle, and above all else the religious tradition that enlightenment thinkers sidelined in favor of naturalism. If human dignity, Havers argues, is to remain the touchstone in the moral considerations challenged by the biogenetic revolution, natural rightists ought to tap the "leavening force of biblical morality."

On a related note Chris Vasillopulos examines the science-based justification for taking life using the Third Reich's policies as a paradigm case and employing Baruch Spinoza as a philosophical backdrop. "Science-Based Justifications in the Third Reich" is a multi-layered argument against the credibility of scientifically informed social policy. The Nazis present the paradigm case since they couched their monstrous Judeocide on dubious hygienic and racial theories which were equated with scientific knowledge. While the Nazis might have turned to the criminalization of everything Jewish, they chose instead to turn Jews into "enemy combatants" that threatened the integrity of a way of life. Vasillopulos shows that some Nazi eradication programs were based on conventional principles of utility and economic calculations, but by the time the Third Reich turned in earnest to the Final Solution their propaganda had so conflated science and values, that the Jew, as a metaphor for disease, had transformed into the Jew as a terroristic threat to an organically defined nation.

On an editorial note, I provide a look at some recent political history in the United States that also highlights the problems associated with the relationship between science and social policy. "Do the Facts Matter? The Politicization of Science" demonstrates how religious values eclipsed the role of scientific advising in recent presidential administrations. This conflict and confounding of the two distinct areas, which are distinguished by different forms of assent and separate goals, has a long history that can be traced to Galileo's problem with the Church. Modern cases, as with Robert Oppenheimer's falling out with the U.S. government and the more recent censorship of Jim Hansen, reflect the same tension that Galileo encountered when his worldview came into conflict with Church doctrine. Oppenheimer's status was undermined because of his leftist history, Hansen fell victim to the claims of scientific uncertainty propounded by business forces and Galileo added fuel to the fire with his superior exegetical skills. Whatever conservative forces threaten scientific revelations, the facts have a way of speaking for themselves.

Tom Winpenny's essay looks at a famous subculture in the United States and examines the stability of its core values in light of recent economic pressures. The culture of the Amish of Lancaster County, the focus of Winpenny's article, is visibly distinct from the urban sprawl and metropolitan development that surrounds them. While the old order may be threatened, particularly as land is not as available as before, their lifestyle is a well-known tourist attraction in Lancaster County and elsewhere. The large billboards displaying caricatured Amish conceal the recent economic pressures that have forced them to consider other means of employment; employment that places them more in the company of the English and which challenges a religiously grounded order that provides regulators that determine if a certain technology is acceptable. The acquisition of wealth is not necessarily a problem but indices like connections to sacred symbols, influence from modern life, exultation of individual achievement, and threats to family integrity are critical. The Amish have adjusted to certain levels of modern technology over the years, but they have been vigilant not to let the *Ordung* erode. The new economic pressures on the Amish of Lancaster County, Winpenny surmises, might be unprecedented and he imagines if the future of the Amish might mean assimilation as it has for the Brethrern and the Mennonites who share their Anabaptist legacy.

The political and environmental fate of a marginalized culture is the subject of Kyle Powys Whyte's essay "Technology, Tribes, and Environmental Racism: From Techno-Oppression to Tribal Sovereignty." If there is a case which bolsters Rousseau's original skepticism regarding the salutary value of technology, the extent to which Indian tribes in North America have been negatively affected by risky technologies could easily qualify. Subject to what Whyte and others call "techno-oppression," the process whereby dominant nations like the U.S. sanction the implementation of risky technologies on or near tribal lands, indigenous populations continue to suffer from the vestiges of institutional racism that originally disenfranchised them. Typical of the environmental injustice which affects economically vulnerable communities, techno-oppression is yet another expression of institutional attitudes which dispossess marginalized groups. Through a case study involving the Isleta Pueblo effort to uphold the standards of the Clear Water Act in order to protect tribal waters used for religious and life-supporting purposes, Whyte explores the viability of a political approach aimed at countering this kind of environmental racism. This legal case successfully invoked U.S. government sanctioned "treatment as a state" status (TAS status) in order to challenge the substandard municipal discharges entering tribal waters upstream. While this approach claims a degree of political autonomy and a kind of sovereignty, it does not come without drawbacks. Whyte argues that TAS status commits an error that should be avoided at all costs in the tribal context because it does not respect the inherent sovereignty of tribal perspectives on environmental risk and its management. In defending this argument Whyte is encouraging philosophers of science and technology and "science and technology studies" practitioners to turn their attention to exploring the possibility of a theory of tribal sovereignty that (1) values tribal interpretations of technology risk and its management and (2) remains binding on state and federal governments.

Tom Easton's concluding essay "A Recession in the Economy of Trust" alerts us to a potential downside to a cutting edge technology, one that promises to be available to PC users as prices come down. For the moment 3D printing, which is essentially a method of fabrication based on compiling slices of print generated materials, is only available to businesses that can afford the $40,000 hardware. Among other things Easton anticipates a Frankensteinian medical use that will eventually generate body parts, but that is not his primary concern. As information has become more and more electronically mediated, software has facilitated methods that can alter and edit original materials. The result can mean undermining trust; trust that we conventionally place in information that is more directly delivered. Easton acknowledges the elusiveness of raw information, unfiltered and undistorted by relative perspectives, but the electronic media has literally placed the power of editing information at our fingertips. Sounds and images can be imperceptibly altered; sometime for good, sometimes to deceive. These trends threaten the social glue of trust and Easton sees on the horizon an even greater risk to the economy of trust from 3D printing. Since we exchange trust in the same way and at the same time economic exchanges occur, 3D printing will introduce a dual threat to the social fabric. But lying about paying back debts and lying about the description of a commodity has always been a perennial problem. As the electronic environment mediates more and more of our lives, it has been necessary to build in certain checks and balances to support trust. Ebay is a case in point. The potential menace presented by 3D printing, Easton claims, far outstrips previous dangers that our trust has faced. To keep pace perhaps we will need to fabricate more effective electronic policing mechanisms that maintain order and cultivate an environment of trust.

Information Ethics: A Critical Assessment

John Barker

The modern revolution in information technology has led us to place more and more value on information. We buy and sell information, a transaction that used to be limited to the exchange of hard, physical objects. We manage and protect online identities, even create whole new ones. We worry about the financial and medical information that is managed on our behalf. In short, information is becoming increasingly consequential to us, leading us to place more and more value on it.

Yet the value we ascribe to information is largely *instrumental* value. We value information because of the impact it has on our lives. We value digital music because we enjoy listening to it. We value the information that makes up our online identities because of the harm that would result if that information were hijacked. In short, we value information for its usefulness. By contrast, we ascribe intrinsic value to ourselves and to other people. If you listen to a digitally downloaded song and enjoy it, the information the song represents is a means to an end for you, but your enjoyment is an end in itself. Most people would accept that the welfare of (at least some) animals can also be an end in itself, making animals valuable intrinsically, not just instrumentally. In short, human beings and (at least some) animals are *moral patients*. A moral patient is anything that has inherent worth and that thereby demands some level of consideration from moral agents.

There is a great deal of controversy about what is and is not included within the category of moral patients. As already noted, most people would now regard some nonhuman animals as moral patients, though this was not always the case. And some environmental ethicists have argued that non-animal species, or entire ecosystems, or even inanimate natural formations, have inherent moral worth and deserve consideration on their own behalf.

Luciano Floridi has recently argued[1] that there is no comfortable place to draw the line between moral patients and moral non-patients. In particular, he argues that informational objects should be recognized as possessors of inherent worth. In other words, the dichotomy described above, between the instrumental value of information and the intrinsic value of those who use information, is a false one on Floridi's view. In fact, Floridi regards *any* attempt to divide up the world in to moral patients and non-patients as hopeless: for Floridi, *being* is an intrinsic moral value, and indeed, it is the most fundamental moral value.

Floridi's view, which he terms "Information Ethics," is one of the most original ideas in contemporary moral theory. But original though it may be, I am going to argue that it faces substantial challenges. I will argue first of all that the main argument for Information Ethics relies on a questionable notion of impartiality. And second, the notion of an informational object is, I will maintain, a rather elusive one. There are innumerable different ways of describing reality in information-theoretic terms, no one of which is better than any of the others. Which one we adopt is largely a matter of convention. This would seem to make ethics largely a matter of convention as well, which is surely an unintended consequence of Information Ethics. Or so I will argue[2]; first, though, we need to get clearer on what Information Ethics is, and what it isn't.

Informational Objects as Moral Patients

Floridi's moral theory is unconventional, to say the least. However, this fact is not, in and of itself, a strong reason to reject it. Consider the idea that animals can be moral patients. That idea must once have seemed very implausible, yet today it is largely for granted. We now mostly agree that at least some non-human animals are entitled to at least some level of moral consideration. In particular, it is now widely held that we ought not to cause animals to suffer needlessly.

The most commonly cited reason for regarding animals as moral patients is their capacity for suffering.[3] Suffering, it is widely felt, is a bad thing wherever it occurs. If this is so, then we have a moral reason not to cause animals to suffer. Depending on the circumstances, this moral reason might be overridden by other considerations; the point, however, is that it would take other considerations to override it. Now Floridi cites animal rights as a precedent for information ethics: if we extend moral consideration to animals, then it is natural to further extend such consideration to other entities, including informational entities. However, there is a weakness in this argument. Namely, unlike animals, informational objects apparently do not experience happiness or suffering.

Now in fact, it is not entirely out of the question that informational objects could experience happiness or suffering. In fact, if a certain view in the philosophy of mind known as Functionalism[4] is true, then a suffering informational object is quite possible. Barring such views as Cartesian dualism, our mental states, including subjective states such as pleasure and pain, depend on our brain states. Now given enough resources, it is possible in principle to create a completely faithful computer simulation of a living, thinking human brain.

In some sense, the simulated brain would think, exactly as the real brain does. According to Functionalism, and indeed probably to the majority of philosophers of mind, the simulation thinks not just "in some sense" but literally, and indeed has all the same mental states as the real physical brain. After all, the brain and the simulation are the same in every way that could matter; they are simply made of different things. And if the simulation and brain really do have *all* the same mental states, then they have the same subjective experiences as well, including the experiences of pleasure and displeasure. If all this is right, then the simulation, an informational object to be sure, has the capacity to suffer, and is therefore, arguably, a moral patient. Thus, there is a strong case to be made for the claim that some informational objects, at least in theory, are moral patients.

However, the argument just sketched is emphatically not Floridi's argument. The case he makes is much more radical. The above argument is based on a fairly conventional view of what it takes to be a moral patient: it takes a capacity for pleasure and displeasure. Floridi bases his moral theory on an entirely different foundation. The most fundamental moral value, for Floridi, is not happiness, nor is it life: the fundamental moral value is *being*, and the most fundamental disvalue is being's opposite, *entropy*. In short, everything in existence has some intrinsic moral worth—possibly very little, or possibly a great deal—simply by existing.

To understand this better, we need to get clearer on what Floridi means by the terms "entropy," "information" and "informational object." Entropy, first of all, is not the entropy of thermodynamics. Floridi defines entropy to be "any kind of *destruction, corruption, pollution,* and *depletion* of informational objects (mind, not of information as content), that is, any form of impoverishment of *being*" (Floridi 2007, pp. 8-9; Floridi's italics). Exactly what this amounts to is a tricky question which I will take up later. For now, think of entropy as the destruction of informational objects.

In the passage just quoted, Floridi contrasts informational objects with information "as content." This is a fairly standard distinction in the theory of information. Consider a piece of information, say a PDF file. There are two ways of approaching the question of the information content of this file. On one approach, one simply asks how much information the file contains: how many bits, say, and perhaps also the values (0 or 1) of these bits. Alternatively, one may ask about the *semantic* contents of the file. The file may, for example, represent an instruction manual for a vacuum cleaner; in pointing this out, we are specifying the file's information contents in the second, semantic sense of "information," not in the first, non-semantic or amount-of-information sense. On the other hand, the file's length in bytes is relevant to its information content in the non-semantic sense of "information," but not in the semantic sense. Floridi uses "information" in the first, non-semantic sense, at least in connection with

Information Ethics (though he has also written about semantic information in other connections[5]).

Finally, we need to understand what Floridi means by an "informational object," as opposed to a mere chunk of information. Here, Floridi's account is complicated, and closely connected to his notion of a *level of abstraction* (LoA).[6] In its most basic terms, a LoA is a way of describing objects or systems that includes some details about the objects described while abstracting away from other details. Or better, a LoA is a mathematical model that can be realized by concrete entities. A feature of reality is described by a LoA by corresponding to an element in the latter's mathematical machinery; those features of reality that we wish to ignore are simply left out of the LoA. (For an exact definition, see Floridi 2008b.) Finally, an object gets to be an informational object by being described as such at an appropriate level of abstraction.

In other words, any object at all can count as an informational object, provided that it is described as such in some LoA. Basically, in specifying an informational object, we specify a set of possible states of the object, together with some rule or set of rules for describing how the object's state changes in response to various conditions. In this way we describe the informational object in the abstract. We describe concrete, particular informational objects by specifying how this abstract abstraction is realized physically: by, for example, specifying what physical states will realize the object's abstract states. For example, computer scientists often describe computer programs in the abstract, specifying the program's possible states, and giving a rule for how the program's state evolves over time. But an abstract computer program can't actually do anything until it is realized concretely, that is, until someone decides to run it. When we run a program on a specific physical machine, we are implicitly making a correspondence between the abstract states of the program, and the physical states of the hardware that realize those abstract states.

The upshot of all this, for our purposes, is simply that just about any object can be regarded as an informational object, provided that it is described as such by some suitable LoA. We simply need to cook up some suitable set of abstract states, etc., to correspond to the object's physical states, etc. Computers, cell phones, slide rules, abaci, etc., are all informational objects, or systems of informational objects, in a fairly obvious way. But more interestingly, even if something is not customarily treated as an informational object, we can regard it as an informational object nonetheless, provided we can find an appropriate LoA. Show me a few dozen marbles, for example, and I'll show you an abacus; the marbles constitute an abacus, and therefore an informational object, as long as they can be *used* as an abacus, even if no one has ever thought to use them as such. Or to use a well known example from the philosopher John Searle,[7] we can regard the stomach as an information processing device if we like; we just have to find some way to interpret the stomach's inputs and outputs as information-bearing. But the latter task is easy, because *everything* is information bearing.

Putting together these two aspect's of Floridi's account—entropy, or destruction of informational objects, as the fundamental moral evil, and the very broad understanding of informational objects via levels of abstraction—we see that for Floridi, just about any object can be viewed as an informational object, and thus as a moral patient. Floridi embraces this consequence of his view, maintaining that even "stars and stones" have some measure of moral worth, however minimal that might turn out to be. Thus, we can now see a little more clearly what Information Ethics is. It is in fact a theory of *arbitrary objects* as moral patients; to regard something as a moral patient, one need merely regard it as an informational object by choosing the right LoA. All it really takes to be a moral patient, and thus to demand some level of consideration from moral agents, is to exist.

Thus, Information Ethics is an extremely ambitious moral theory. Instead of applying a more conventional moral framework to the special case of information, it seeks to rewrite moral theory itself in information-theoretic terms. As such, it carries with it a rather strong burden of proof, a point to which I now turn.

Motivation

As we have seen, Information Ethics holds that all objects in existence possess some measure of intrinsic moral worth, and that existence is the one fundamental moral value. Floridi's main argument for this claim is based on the notion of *impartiality*. It is widely felt that ethics must be impartial: it must treat everyone equally. For example, Utilitarianism judges an act in terms of the net amount of happiness it produces (with unhappiness treated as negative happiness). In so doing, it treats one person's happiness as just as important as anyone else's, no more and no less. Utilitarianism therefore embodies a certain kind of impartiality, and this impartiality is often cited as a strength of Utilitarianism. Impartiality also motivates the recognition of animals as moral patients: the alternative, recognizing only human beings as moral patients, discriminates against animals without justification. Floridi argues that Information Ethics simply takes impartiality to its logical conclusion, treating all of existence equally.[8]

Moreover, Floridi argues compellingly that the history of ethics involves a steady movement toward greater impartiality. Before people started thinking systematically about ethics, they withheld the status of moral patient from all but the members of their own tribe or nation. Later, this status was extended to the whole of humanity. Many if not most people would now treat at least some non-human animals as moral patients. Some would ascribe moral worth to entire ecosystems and even to inanimate parts of nature. (This view, a school of environmental ethics known as land ethics, regards natural formations as inherently valuable, for their own sake and not simply because human beings happen to take an interest in them.) Thus, the history of ethical thinking is one of successively widening the sphere of our moral concern, and the logical end result of this process is to extend our moral concern to all of existence—or so Floridi argues.

I would maintain, however, that this argument relies on a questionable notion of impartiality. Being impartial means treating the targets of one's actions equally; but "equally" is notoriously hard to define in this context. For example, we all agree that systems of criminal justice should be impartial. Yet a good system of criminal justice always discriminates to this extent: it treats the guilty (or the convicted) differently than it treats the innocent (or the acquitted). When we say that a system of justice is impartial, we mean that everyone who lives under its sway is treated according to the same set of rules, with no special exceptions for anyone. But the rules themselves discriminate in favor of some types of behavior and against others and rightly so.

Thus, impartiality is not the same thing as nondiscrimination. Instead, impartiality demands that when we do discriminate, we do it for the right reasons. It demands that the basic principles of morality apply equally to everyone, but it does not dictate the form that those principles must take. Of course, to achieve true impartiality, it is not quite enough that the same rules be applied to everyone; it is also necessary that the rules themselves be impartial. For example, consider the rule: "Never accept a bribe, unless you live in Chicago." Such a rule would be problematic from an ethical standpoint, even if it were applied equally to everyone, because the rule itself fails to be impartial: it treats Chicagoans differently from everyone else. By contrast, the rule "Never accept a bribe" does seem to be impartial, even though it treats bribes differently from other transactions. In general, then, impartiality demands that everyone be treated equally according to rules that are themselves impartial. The fact that "impartial" appears in its own definition makes the entire topic of impartiality a tricky one indeed.

Now consider again Floridi's argument that ethics has been moving toward greater impartiality, with Information Ethics as the natural end result of this movement. As one stage in this progression, consider again the Utilitarian view that we should always act so as to maximize the world's net supply of happiness. This principle seems impartial, because it treats everyone's happiness equally. But if we understand this principle as a call to specifically increase *human* happiness, then arguably the principle fails to be impartial, since it treats animals differently from humans without any clear justification. For this reason, many if not most Utilitarians regard animal happiness as inherently valuable and thus recognize animals as moral patients. Now some would argue that the maximum happiness principle is still too narrow, in that it discriminates against conscious beings and other living beings, e.g., plants. This leads naturally to the idea of life as a fundamental moral value. Still others feel that even this is not impartial enough, as it discriminates against nonliving but still valuable parts of nature. And finally, Floridi holds that anything short of regarding existence itself as the one fundamental moral value constitutes an unjustified form of discrimination in favor of some parts of the universe and against others.

However, it is by no means clear that this line of argument is sound. As we have seen, impartiality does not demand a complete lack of discrimination. It simply demands a lack of unjustified discrimination. If, for example, we adopt Utilitarianism's maximum happiness principle, we are indeed discriminating against those beings that lack a capacity for happiness and unhappiness. But this may be a fully justified form of discrimination. It simply reflects the proposition that happiness has more inherent value than life. Any moral theory must view some outcomes as having more positive moral value than others: otherwise there would be nothing to choose between different outcomes, and little point to moral decision making. Thus, any moral theory must discriminate in some respect. And Information Ethics is no exception, since it regards some objects as having "minimal" and "overridable" moral value, while acknowledging that other objects have greater value.

Thus, by itself the impartiality argument seems rather too weak to motivate Information Ethics. Any moral theory will have to decide what features of objects or outcomes are morally relevant and which features (e.g., residing in Chicago) are not; the mere fact of making such a discrimination does not amount to impartiality.

Application

Assuming, however, that we accept being and entropy as the fundamental moral good and evil, respectively, the question remains how to translate this idea into moral judgments about particular cases. And it is here that I think Information Ethics faces its most serious challenge.

Let's start with the idea of existence as the one fundamental moral good. If we assume that we are to act in a way that leads to the most good (as compared to the alternative actions we might perform), then Information Ethics would seem to require us to act in a way that maximizes the number of objects in existence—at least if we take "existence" literally. However, such a prescription would be problematic for a number of reasons. For one thing, *any* action creates countless objects and destroys countless others. Remember, the relevant notion of "object" here is a very broad one. It includes not just familiar objects like tables and chairs, but also informational objects like web pages and avatars. It includes the individual CO_2 molecules that we create every time we draw breath, and the individual O_2 molecules that we destroy in the same process. And it presumably even includes such heterogeneous and scattered objects as the top half of the Eiffel Tower, or every third wheel of a fleet of trucks. It is probably impossible to count the number of objects that are created or destroyed by a given action; but even if we could, such a count would surely bear little or no relation to that action's rightness or wrongness.

It seems that Floridi would agree with this assessment, since he maintains that not all objects are created equal, morally speaking. Some objects, that is, have a great deal of intrinsic moral value, while others have very little moral value. In addition, it is not the mere existence of objects that is morally relevant, apparently, but the flourishing of such objects as well. In other words, we need interpret "being" and "entropy" more broadly in the context of Information Ethics than I did in the last paragraph.

Information Ethics fundamentally calls on us to avoid entropy: to avoid creating it and to minimize it where it already exists. But what exactly is meant by "entropy" here? As we have seen, Floridi defines entropy as "any kind of *destruction*, *corruption*, *pollution*, and *depletion* of informational objects (mind, not of information as content), that is, any form of impoverishment of *being*" (Floridi 2007). Unfortunately, as a guide to action, this definition leaves something to be desired. Suppose we find ourselves in a situation in which no matter how we act, we will wind up destroying, corrupting, etc., at least some informational objects. Indeed, as I indicated earlier, it is likely that any act will create some informational objects and destroy others, and so it is likely that any act whatsoever will lead to some amount of corruption, pollution, and depletion of informational objects as well. In that case, given two actions that would both lead to some amount of destruction, corruption, etc. of informational objects, the question naturally arises: which act should be performed? Which act would lead to less entropy? Obviously, the question is meaningless unless we have some way of quantifying entropy. However, the above definition of entropy is qualitative, not quantitative. It describes the kinds of things that count as entropy, but it does not specify *how much* entropy is generated in a given instance of destruction, depletion, etc. Nor does it specify what it is for one instance of destruction, etc. to constitute more entropy than another. Without some quantitative, or at least comparative, measure of entropy, Information Ethics simply fails to yield any moral recommendations in particular cases.

Now one fairly straightforward way to define entropy quantitatively is to simply equate it with thermodynamic entropy. Roughly speaking, thermodynamic entropy measures the amount of order or disorder in a given physical system, with higher entropy indicating greater disorder. If we identify entropy with thermodynamic entropy, then Information Ethics becomes a call to maximize the amount of order and structure in the universe. However, there are two good

reasons to reject such an identification. First, Floridi himself rejects this identification quite explicitly. And second, it would cause Information Ethics to run
afoul of the Second Law of Thermodynamics. According to the Second Law,
the entropy of an isolated physical system always increases over time. Since
the universe itself is an isolated system, the total thermodynamic entropy of the
universe is an ever-increasing quantity. Thus, any attempt to minimize, or even
simply to decrease, the amount of entropy about is doomed to failure.

There is a partial loophole in the Second Law: entropy can decrease in one
localized part of the universe, provided that it increases somewhere else. For
example, in building a house, a relatively high-entropy bit of matter (in this case
a pile of lumber) is put into a configuration (a house) with lower entropy. But
this is no violation of the Second Law, because the house/lumber in question is
not a closed system. Building a house takes energy: to build a house we need
to burn fuel, and when we burn fuel, we thereby take it from a state of lower
entropy to a state of higher entropy. So in building a house, we are trading lower
entropy in one place for higher entropy elsewhere.

Now it may well happen that we can decrease the entropy of a part of the
world that we care about, at the expense of higher entropy in a part of the world
that we care less about. House construction is one of many examples of this
phenomenon. Thus, we might try to save the thermodynamic interpretation of
Information Ethics by construing it as a call to decrease thermodynamic entropy locally, in the parts of the world we care most about. However, this runs
against the spirit of Information Ethics in an important way. As we have seen,
the primary motivation for Information Ethics is a desire to liberate ethics from
an anthropocentric viewpoint. Informational objects are to be viewed as morally
worthy on their own account, not just instrumentally valuable because of the
interest we happen to take in them. Moral value is to be divorced from specifically human concerns. This idea would be completely undermined if entropy,
the fundamental measure of moral disvalue, were defined in terms of the adverse
effects our actions have on the parts of the world we care most about, while
ignoring their effects on the parts we care least about. Surely, how much we
care about an object should be irrelevant to that object's intrinsic moral worth
if the spirit of Information Ethics is to be preserved at all.

In the context of Information Ethics, a good notion of entropy should do two
things. First, it should be specific enough to generate concrete moral recommendations in specific cases, at least in principle. Moreover, these recommendations
should not be wildly at odds with our ordinary moral judgments; they should
not, for example, tell us that killing a computer file is more seriously wrong
than killing a human being.[9] Second, a good notion of entropy should support
Information Ethics' aspiration to be a non-anthropocentric moral theory. If, for
example, we define the entropy of an object to be some measure of how much
we care about it or of how much instrumental value it has given our interests,

then such a notion would be an unsuitable foundation for Information Ethics, even if it yields all the right moral judgments.

In the next two sections, I will examine two candidate definitions of entropy. The first definition is drawn from Claude Shannon's theory of statistical information; the second is based on the theory of computational complexity. In each case, I will argue that while the given theory of information is sound, its application to ethics is dubious. The discussion will of necessity be somewhat technical. In the final section, I develop some of the same ideas in a less technical setting.

Entropy via Shannon Information Theory

As we have seen, if we are to take seriously the idea that "being" is the most fundamental good and "nonbeing" or "entropy" is the most fundamental evil, we cannot calculate good or evil by simply counting objects. A natural idea, and one which is somewhat suggested by Floridi's term "entropy," is that fundamental moral value should be identified with some overall measure of informational richness or complexity. This would preserve the idea of being and nonbeing as fundamental moral values while avoiding the difficulties involved in the simple counting approach.

One of the best-developed accounts of non-semantic information is *statistical information theory*. This theory, developed by Claude Shannon in the 1940s,[10] has been used very successfully to describe the amount of information in a signal without describing the signal's semantic content (if any). Thus, it seems like a natural starting point for describing the overall complexity or richness of a system of informational objects.

Statistical information theory essentially identifies high information content with low probability. Specifically, the Shannon information content of an individual message M is defined to be $\log_2(1/p(M))$, where $p(M)$ is the probability that M occurs.[11] As a special case, consider a set of 2^n messages, each equally likely to occur; then each message will have a probability of 2^{-n}, and an information content of $\log_2(2^n) = n$ bits, exactly as one would expect. The interesting case occurs when the probability distribution is non-uniform; low probability events occur relatively rarely, and thus convey more information when they do occur.

As is well known, the definition of Shannon information content is formally almost identical to that of thermodynamic entropy. The statistical entropy S of a given physical system is defined to be $S = k_B \ln \Omega$, where k_B is a constant (Boltzmann's constant) and Ω is the number of *microstates* corresponding to the system's *macrostate*. (A system's macrostate is simply its macroscopic configuration, abstracting away from microscopic details; the corresponding microstates are those microscopic configurations that would produce that macrostate.) Now for a given microstate q and corresponding macrostate Q, Ω is simply the probability that the system is in microstate q given that it is in macrostate Q. In other words, the entropy of a system is simply $k_B \ln (1/p_Q(q))$, where p_Q is a uniform probability distribution over the microstates in Q. Alternatively, if we posit a uniform probability distribution p over all possible microstates q, then we have $p_Q(q) = p(q) / p(Q)$, and thus $S = (k_B/p(q)) \ln p(Q) = -(k_B/p(q)) \ln (1/p(Q))$; the quantity $k_B/p(q)$ is a constant because the measure p is uniform. In any case, we have $S = K \log (1/p)$, where K is a constant and p is the probability of the state in question under some probability measure (the base may be omitted on the log because it only affects the result up to a constant, and may thus be subsumed in K). Thus, up to a proportionality constant, statistical entropy is a special case of Shannon information content.

However, it is the *wrong* special case, as we have already seen. Thermodynamic entropy is simply not a good candidate measure of moral evil. Thus, if we are to use Shannon information theory to capture the morally relevant notion of complexity, we will have to use a probability measure other than that described above. However, information theory does not offer us any guidance here, because it does not specify a probability measure: it simply *assumes* some measure as given. Typically, when applying information theory, we are working with a family of messages with well-defined statistics; thus, a suitable p is supplied by the context of the problem at hand.

Thus, Shannon information theory provides a measure of a system's information content, but this measure is relative to a probability measure p. This presents an obstacle to explaining complexity in terms of Shannon information and simultaneously claiming that complexity is a fundamental, intrinsic moral value. If we allow complexity to be relative to a probability measure, then intrinsic moral worth will also be relative to a probability measure. Conceivably, different probability measures could yield wildly different measures of complexity and, thus, of intrinsic moral worth. Thus, it would appear to be necessary to pin down a single probability measure, or at least a family of similar probability measures, in a non-arbitrary manner.

And here is where things get tricky. What probability measure is the right one for measuring the complexity of *arbitrary* systems? Whatever it is, it must be a probability measure that is in some sense picked out by nature, rather than by our own human interests and concerns. Otherwise complexity, and thus inherent moral worth, is not really objective, but is tied to a specifically human viewpoint. This goes against the whole thrust of Information Ethics, which seeks to liberate ethics from an anthropocentric viewpoint. Thus, we need to find a *natural* probability measure for our task. What might such a probability measure look like?

The best-known conception of objective probability is the frequentist conception. According to that conception, the probability of an outcome O of an experiment E is the proportion of times that O occurs in an ideal run of trials of E. To apply this notion, we need a well-defined outcome-type O, a well-defined experiment-type E, and a well-defined set of ideal trials of E –and if the latter set is continuous, a well-defined measure on that set. This is all notoriously difficult to apply to non-repeatable event tokens and to particulars in general. To assign a frequentist probability to a particular x, it is necessary to subsume x under some general type T, and different choices of T may yield different probabilities. In other words, the frequentist probability of a particular depends among other things on how that particular is described. Different ways of describing a particular will correspond to different conceptions of what it is to repeat that particular, and thus, to different measures of how frequently it occurs in a run of cases.

What this means for us is that the information content of a concrete particular depends, potentially, on how we choose to carve up the world. Again, this is not a problem in practice for information theory, since in any given application, a particular (frequentist) probability measure is likely to be singled out by the problem's context. But in describing the information context of completely arbitrary objects, there is no context to guide us. In particular, if we subsume a concrete particular x under a commonly occurring type T, it receives a high frequentist probability, and correspondingly low Shannon information content. If we subsume that same particular under a rarely occurring type T^*, it receives a low probability and correspondingly high information content.

Thus, it is by no means obvious that there is a choice of probability measure that (a) is natural independently of our own anthropocentric interests and concerns, and (b) gives us a measure of complexity that is a plausible candidate for inherent moral worth, even assuming that the latter has any special tie to complexity in the first place. To be fair, it is also not obvious that there is not such a probability measure. As the measure p from thermodynamics shows, there is at least one natural way of assigning probabilities to physical states, one which does indeed yield a measure of complexity, albeit not the measure of complexity we are looking for. It also raises a further worry. The reason thermodynamic entropy is a bad candidate for basic moral disvalue is simply that it is always increasing, regardless of our actions. That is simply the Second Law of Thermodynamics. What guarantee do we have that complexity, measured in any other way, is not also decreasing inexorably? Thermodynamic entropy can decrease *locally*, in the region of the universe we care about, at the expense of increased entropy somewhere else, and the same may be true for other measures of complexity. But this fact is surely irrelevant to a patient-centered, non-anthropocentric moral theory.

Entropy via Computational Complexity

In light of the preceding section, Floridi can respond that his notion of entropy is simply not that of Shannon. And this response is correct, as far as it goes.

However, it does raise two important issues. First, it raises the question of what measure of entropy should be used, if not the Shannon measure. We have seen that Floridi defines entropy as "any kind of *destruction, corruption, pollution, and depletion* of informational objects," but we have also seen that this informal characterization of entropy is simply not specific enough to serve as a foundation for Information Ethics. If we also reject the Shannon definition of entropy, then we have a right to ask what definition of entropy should be used.

More importantly, the deficiencies of the Shannon approach are actually quite general, I would argue, and do not depend on the specifics of Shannon's theory. To isolate these general issues, let us briefly examine an alternative to Shannon information theory, the theory of computational complexity.

One way to measure the complexity of a problem is to measure the amount of computer time, memory or code that it would take a computer to solve the problem. This idea is the foundation of an important branch of theoretical computer science known as *computational complexity theory* (sometimes known simply as complexity theory). An interesting variation on complexity theory measures the complexity of an object in terms of the amount of computer code that it would take to generate that object. Specifically, Gregory Chaitin defines[12] the complexity of a string of symbols to be the length of the shortest program (in a certain specified computer language) that would generate that string as output. Chaitin even shows that with a suitable choice of probability measure, his definition of information content coincides with that of Shannon.

Now using Chaitin's complexity measure, we can certainly assign a well-defined complexity $H(s)$ to any given string s of symbols. We can even define the entropy of s in terms of $H(s)$: say, entropy of $s = 1/H(s)$. And limiting ourselves to strings of symbols is not too severe a restriction, when one realizes that any data at all can be represented as a string of symbols: after all, a file or data stream is nothing more than a sequence of 0s and 1s, that is, a string in the two-letter alphabet consisting of 0 and 1.

However, the strings of symbols to which Chaitin's measure applies are abstract objects, not concrete particulars. Take the five-letter string "Hello," for example. This string can appear in many places at once: it appears in every copy of this work, is written down on countless scraps of paper, etc. In short, "Hello" is an abstract type that has many concrete instances, or tokens. As another example to illustrate the distinction, the play *Hamlet* is an abstract type, while any individual printed copy of the play is a token. Types are mathematical objects, and it is to types that Chaitin's (or for that matter Shannon's) measure of complexity applies directly. When we speak of a particular token T having a certain Chaitin complexity, we simply mean that T is a token of some type that has that Chaitin complexity.

On the other hand, it is tokens, not types, that are the best candidates for moral patienthood. If a given token of "Hello" has inherent moral worth, that means that morally speaking, it matters what we do to that token. Destroying that token might be morally less preferable than preserving it, for example. But there is no such thing as destroying or preserving a type. Indeed, there is no such thing as doing something to, or having an effect on, a type. Types are abstract objects, and are therefore not the sorts of things that can be acted upon. In a similar vein, a dozen eggs might constitute a moral patient, but the number 12 itself is not a moral patient. A dozen eggs can be discarded or made into a soufflé, but neither thing can happen to the number 12.

Thus, to base Information Ethics on computational complexity, we need a measure of complexity that applies not just to types, but to tokens as well. Now defining such a measure may seem completely trivial. Namely, if s is a string type and T is one of its tokens, then the complexity $H(T)$ of T should be defined simply as the complexity $H(s)$ of the corresponding type. This makes perfect sense, as far as it goes. What it ignores, however, is the fact that a given physical object may be regarded as a token of many different types, and that which type a given token represents is, fundamentally, a matter of convention.

Take the string "Hello" again, for example. By convention, we associate certain letter shapes with certain abstract letter types; this convention is more complicated that it may seem, since it has to account for different fonts and typefaces. We also, by convention, read and write strings left-to-right; under a different convention, the same physical inscription would represent the string-type "olleH." We have also decided, by convention, that the same shape represents the same symbol regardless of where it occurs in a string-token. Some writing systems adopt a different convention; in Arabic, for example, a letter can take a completely different form depending on whether it occurs at the start, in the middle, or at the end of a word. We could have taken a similar approach, and decided that the two occurrences of "l" in "Hello" represent different letter-types because they occur in different positions.

In general, we can adopt any convention we please. We can, for example, devise a complicated encryption algorithm that maps non-encrypted strings s to encrypted strings $F(s)$; having done that, we can stipulate that if a given inscription is a token of a string s under the "standard" convention, then it is a token of the string $F(s)$ under the new convention. The function F is pretty much arbitrary here. In particular, it need not preserve Chaitin complexity or any other measure of complexity: s and $F(s)$ may have quite different complexities. Thus, the convention we adopt for interpreting inscriptions can have an effect on the complexity we assign to the various inscriptions, even if we have settled on a single measure of complexity for abstract strings.

And here is where we run into the same problem we encountered with Shannon information theory. Of all the possible conventions we could have adopted for interpreting concrete inscriptions, we have settled on a convention that is convenient for us. Since a physical inscription's complexity, and thus its entropy, is convention-dependent, so is its moral worth if we identify the latter with informational complexity. And this is bad news for Information Ethics, since it runs directly counter to the latter's goal of defining moral worth independently of human interests and concerns.

This point is easy to miss, because when we think about a physical object as a bearer of information, we have already adopted a convention that lets us speak unambiguously of that object's information content. Thus, the issues I am raising here may seem not to arise at all. However, if we regard an object simply as a physical object, then we are rather at a loss to say what information it does or does not contain. For Floridi, even "stars and stones" can have moral worth, but what is the information content of a stone? Well, we can regard a stone as information-bearing if we so choose. We can regard it as a letter in an alphabet, say. Or we might regard the various parts of the stone as representing different letters, so that the stone represents a word or even a whole sentence. The point is that it is largely up to us to decide how much information there is in a stone. But it should not be up to us to decide how much intrinsic moral worth there is in a stone or in anything else. (Of course, to a great extent it is up to us to decide how much instrumental moral worth there is in a stone or in anything else.)

Information Everywhere

One last example may help clarify this point. How much information is there in a glass of water? The obvious, intuitive answer is: very little. A glass of water is fairly homogeneous and uninteresting. Yet the exact state of a glass or water would represent an enormous amount of information if it were described in its entirety. There are approximately 7.5×10^{24} molecules in an eight ounce glass of water.[13] If each molecule has a distinguishable pair of states, call them A and B, then a glass of water may be regarded as storing over seven trillion terabits of data. Further, let f be any function from the water molecules into the state set $\{A, B\}$. Relative to f, we may regard a given molecule M as representing the binary digit 0 if M is in state $f(M)$, and 1 otherwise. Clearly, there is nothing to prevent us from regarding a glass of water in this way if we so choose, and with any encoding function f we like. And clearly, by a suitable choice of f, we may regard the water as encoding any data we like, up to about seven trillion terabits. For example, by choosing the right encoding function, we may regard the water as storing the entire holdings of the Library of Congress, with plenty of room to spare. Alternatively, a more "natural" coding function, say $f(M) = A$ for all M, might be used, resulting in a relatively uninteresting but still vast body of information.

Now if ordinary objects like glasses of water really do contain this much information, then there is too much information in the world for information content to be a useful measure of moral worth. The information we take a special interest in—the structures that are realized in ways that we pay attention to, the information that is stored in ways that we can readily access—is simply swamped by all the information there is. The moral patients we normally take an interest in are vastly outnumbered by the moral patients we routinely ignore. Thus, if information content is to serve as a measure of moral worth, the vast quantity of information in a glass of water must be excluded.

But on what basis could we exclude it? We might try to exclude some of the more unconventional encoding functions, such as the encoding function that represents the water as storing the entire Library of Congress. Such encoding functions, it may be argued, are rather unnatural and do not represent the information that is objectively present in the water. Even if this is so, there is no getting around the fact that a glass of water represents a vast amount of information, in that it would take much information to accurately describe its complete state. That information might be rather uninteresting—uninteresting *to us*, that is—but so what? If moral worth is tied to information content per se, then it does not matter whether that information is interesting. If moral worth is tied to *interesting* information, then it appears that moral worth is directly tied to human concerns after all.

But there is a more fundamental problem with dismissing some encoding functions f as unnatural. Whenever information is stored in a physical medium, there needs to be an encoding function to relate the medium's physical properties to its informational properties. Often, this function is "natural" in that it relates a natural feature of information (e.g., the value of a binary variable) to a natural feature of the physical medium (e.g., high or low voltage in a circuit, the size and shape of a pit on an optical disk, magnetic field orientation on a magnetic disk, etc.). However, there is absolutely no requirement to use natural encoding functions. There need be no simple relation whatsoever between, say, a file's contents and the physical properties of the media that store the file. The file could be encrypted, fragmented, stored on multiple disks in a RAID, broken up into network packets, etc.

In practice, we always disregard the information that is present, or may be regarded as present via encoding functions, in a glass of water. But the reason does not seem to be a lack of a natural relation between the information and the state of the water. The reason is that even though the information is in some sense there, we cannot easily use or access it. We can regard a glass of water as storing a Library of Congress, but in practice there is no good reason to do so. By contrast, a file stored in a possibly very complicated way is nonetheless accessible and potentially useful to us.

If this is right, then there is a problem with viewing information's intrinsic value as something independent of our own interests as producers and consumers of information. The problem is that information does not *exist* independently of our (or someone's) interests as producers and consumers of information. Or alternatively, information exists in an essentially unlimited number of different ways: what we count as information is only a minute subset of all the information there is. Which of these two cases obtains is largely a matter of viewpoint. On the former view, even if inanimate information has moral value, it has value in a way that is more tied to a human perspective than Floridi lets on. On the latter, there is simply too much information in the world for our actions to have any net effect on it.

Notes

1. See, e.g., Floridi 2007, Floridi 2008a, and references therein.
2. Some of the arguments I will make were also made in my (Barker 2008).
3. The classic source of this argument can be found in the writings of Jeremy Bentham and John Stuart Mill; see, e.g., Mill 1863. The best known contemporary version of the argument is due to Peter Singer; see, e.g., Singer 1975.
4. Functionalism basically holds that the character of a mental state is determined by its interactions with other mental states, and is not essentially tied to the specific physical state that realizes it. In particular, the fact that our own mental states are realized by brain states is not an essential feature of those mental states, which could be realized in a completely different medium, a computer for example.
5. See Floridi 2005.
6. Floridi has developed the concept of levels of abstraction in a number of publications. See in particular Floridi 2008b.
7. See Searle 1980.
8. See in particular section 2.1 of Floridi 2008a.
9. At least, all else being equal: some human beings may have it coming, and some computer files may be so important that their deletion would lead to serious harm. The latter point, however, goes more to the instrumental value of the file than to its intrinsic value.
10. See Shannon (1948). For a good modern introduction see MacKay (2003).
11. A base-2 logarithm is used because information is measured in bits, or base-2 digits. If information is to be measured in base-10 (decimal) digits, then a base-10 logarithm should be used. In general, the Shannon information content is defined to be $\log_b (1/p(M))$, with b determined by the units in which information is measured (bits, decimal digits, etc.).
12. See Chaitin 1990.
13. This figure is obtained from the number of molecules in a mole (viz. Avogadro's number, approximately 6×10^{23}), the number of grams in one mole of water (equal to water's atomic weight, approximately 18), and the number of grams in 8 ounces (about 227).

References

Barker, John. 2008. "Too Much Information: Questioning Information Ethics." *APA Newsletter On Philosophy and Computers* 8.1. 6-19.

Chaitin, Gregory J. 1990. *Algorithmic Information Theory*. Cambridge: Cambridge University Press.

Floridi, Luciano. 2005. "Is Semantic Information Meaningful Data?" *Philosophy and Phenomenological Research* LXX. 351-370.

---. 2007. "Understanding Information Ethics." *APA Newsletter On Philosophy and Computers* 7.1. 3-12.

---. 2008a. "Information Ethics, its Nature and Scope." In *Moral Philosophy and Information Technology*, edited by Jeroen van den Hoven and John Weckert. Cambridge: Cambridge University Press.

---. 2008b. "A Defence of Informational Structural Realism." *Synthese* 161.2. 219-253.

MacKay, David J. C. 2003. *Information Theory, Inference, and Learning Algorithms*. Cambridge: Cambridge University Press.

Mill, John Stuart. 1863. *Utilitarianism*.

Searle, John. 1980. "Minds, Brains, and Programs." *Behavioral and Brain Sciences* 3 (3). 417-457.

Shannon, Claude. 1948. "A Mathematical Theory of Computation." *Bell System Technical Journal* 27. 379-423, 623-656.

Singer, Peter. 1975. *Animal Liberation*. New York: Random House.

Teaching Values in Computing Courses through Theory and Practice

Fani Zlatarova

Introduction

Focusing on the ethical aspects of cyberspace, while teaching students majoring in computing disciplines has become increasingly important as their professional relationships affect large groups of users and have significant influence on those users' activities. There are two different approaches to introducing ethics to these students:

- as a part of the required teaching material in computing courses, and
- through various forms of out-of-class activities.

The latter approach is discussed in this chapter by focusing on variety of student-led projects, which blend ethical and specific computing topics. Other methods which also follow the second approach, such as online surveys, seminar organization, quiz bowls, information bulletin board design, and field trips are also presented. In addition, several academic forums and committees that have been used as arenas to present and discuss cyberethics, the moral aspects of information systems (ISs), and the development of appropriate projects and the organization of related events are described.

In most cases, curricula developed for computing majors do not offer special courses in ethics [2, 3, 8, 9]. That is why some students graduate with limited or no knowledge about the ethical principles relevant to the numerous aspects of computing. A significant number of instructors try to include ethics topics in their classes by introducing some basic principles through different teaching tools [15, 33, 34]. Among various academic institutions, it is possible to distinguish a very specific group of institutions that considers as an academic priority not only the teaching and research activities, but also service learning which can be described as "the educational practice and philosophy of integrating classroom concepts with a related community service experience" [25]. Primarily, these are liberal arts colleges. Strict requirements are imposed on their students and faculty to participate in service-oriented activities and events and to generate appropriate ideas and solutions. Such an atmosphere creates an excellent environment that fosters various out-of-class activities which introduce knowledge relevant to cyberethics issues. The main objective of this paper is to share the author's experience in implementing different forms of teaching ethics to students majoring in computing disciplines at a liberal arts college.

Computer Ethics as a Class Experience

According to Laudon [15], ethics consists of "principles of right and wrong that can be used by individuals acting as free moral agents to make choices to guide their behavior." Life is controlled by a variety of rules and imposing additional rules of behavior meets serious resistance [16]. Teaching ethics in computing classes raises problems for the instructor because sometimes the breaking of commonly accepted principles of ethical behavior relevant to cyberspace is perceived as a demonstration of professionalism and advanced understanding and use of hardware and software [5, 12, 27-28, 31]. For example, how should students be led to believe that they are doing wrong by downloading copyrighted music? So many others are doing it and it is so easy and appealing to do. How can they be stopped from copying ideas and even texts from online sources for their assignment preparation? How should they be taught to surf safely through networks while keeping their privacy?

One possible answer to the above questions lies in interjecting various ethical topics into traditional lectures and into non-traditional forms of teaching, for example, development of research projects, quiz bowls, student electronic presentations, websites, and concrete case studies reflecting students' experiences [17, 19]. Ethical behavior could be considered when creating a program that is used by others, when developing an IS or when designing a website and uploading it on servers accessible through the Internet, when buying books online or researching resources from the WWW for assignment purposes, when downloading texts, pictures, or music files, and so on. These daily routines impose high responsibilities for students as computing professionals and as members of society. Finding different ways of teaching and educating according to the cyberethics principles adds precious knowledge to their academic background. Exploring the service-learning opportunities at liberal arts colleges represents another effective and efficient way for involving students in out-of-class activities related to ethical issues.

A variety of methods have been used to provoke students' interest and to get them involved in related activities such as:

- teaching specific topics;
- discussing case studies;
- writing essays and research papers;
- offering electronic presentations that also include ethical issues;
- creating codes of ethical conduct for virtual companies as a part of corresponding websites and ISs developed by students;
- organizing surveys and quiz bowl sessions;
- participating in group class projects or individual senior projects.

Practical solutions for introducing ethical and moral aspects in different computing courses are described below.

The Information Technology and Society *Course*

According to Spinello [28], "the information age has created a more open society where piracy seems to grow ... with each technological innovation." The Internet has become a mirror of individual behavior. A significant portion of cyberspace users are young people who enthusiastically use all possible opportunities to access and process information.

The *Information Technology and Society* course is offered as a freshman seminar primarily for students who intend to select a major, a minor, or a concentration related to computing. Such students are usually interested in learning more about computer applications. The main goals of the course are to prepare the students to use information technology (IT) in their academic work during their college years and to show them the respective recommended ethical principles that should guide them as members of society when using IT methods and tools.

Students' attention is focused on seven factors [13] that challenge the moral codes of the computerized society: speed, privacy and anonymity, the nature of the medium, aesthetic attraction, increase of potential victims, international scope, and the power to destroy.

Teaching students how to write papers as a part of their academic assignments is an important objective of the instructor's class work. In the case of this course, individual and team papers are oriented towards discussions of basic ethical issues such as intellectual property and rights, privacy and security, and human factors—all relevant to the usage of computers and software development. Special attention is paid to the concept of copyright and corresponding international and U.S. laws. While researching a particular area, students must evaluate a moral dilemma in an unbiased manner, relying on logic and evidence. All students are required to write essays on the following two topics: *"Pirated Software. Do I know it?"* and *"Is My Computer a Health Hazard?"* by including quantitative and qualitative estimations of research elements on related ethical issues.

From the very beginning of the semester, students begin to develop an individual progressive paper by choosing a favorite topic that reflects a moral dilemma relevant to computers. Listed below are several topics chosen by students in the past:

- *Computers based on biodigital technologies: How will they change our society?*
- *Computers and IT—forces of evil.*
- *Computers and the human interaction of tomorrow.*
- *The future intelligent machine society: Do human beings belong to it?*
- *Computers own my life—should I agree with this?*
- *Interactive multimedia: eye, ear, hand, and mind. Does it kill children's imagination and creativity?*
- *The controversy about the first contemporary computer. Who was the father?*
- *Human beings—slaves or leading force in the machines' world of tomorrow?*
- *My unlimited virtual life versus my lonely real life: what is my choice?*
- *Giving up privacy in exchange for safety: Reasons and limitations.*
- *Identity theft: Surreal or real?*
- *Open source software: pros and cons.*

The development of the progressive paper is useful not only for students but also for the instructor. By evaluating them throughout the semester, the lecturer develops an understanding of students' individual perceptions and reasoning and thus leads them into more targeted and focused discussions.

Other forms of teaching and assessment are also used in this course. Some of them are considered below.

Case Studies and Electronic Presentations. Using case studies to analyze and judge real and imaginary situations is helpful because they leave a strong impression and they use concrete examples to illustrate human behavior and its associated consequences. Internet news sources, together with other types of media, offer interesting opportunities for this. After such classes, students are able to consider ethical issues related to personal information uploaded to the Web while searching for online products or services. Usually, the respective case studies involve individual or group electronic presentations.

Team Projects. Another way of teaching ethical behavior is explored by assigning team projects consisting of creating a website for a virtual company and writing a corresponding code of ethics and professional conduct for the employees of this company. The text adopted by the Association for Computing Machinery (ACM) available through the organization's website [36] is a good example for such a code and can be used as a prototype. Another good example is the Code of Ethics for Computing Professionals that was developed by the Institute of Electrical and Electronics Engineers (IEEE) [38].

Quiz Bowls. Quiz Bowl questions that raise ethical problems and provide relevant knowledge are exciting for students. They often prefer such non-traditional classes because of the competitive atmosphere, the opportunity to learn unusual entertaining facts, and the overall contribution to the class activity portion of their final grade.

Class Lectures. Traditional class lectures on selected ethics topics are also presented by the instructor to complete the introduction of ethical concepts and principles in the course content.

The knowledge gained from the *Information Technology and Society* course allows for better understanding of social context while using IT in everyday academic activities and while preparing assignments, regardless of the specific discipline. Students are expected to develop a strong background that becomes useful when considering issues related to:

- decision-making involving concepts such as intellectual property;
- open-source software;
- copyright and corresponding plagiarism problems;
- on-line communication;
- privacy protection and privacy-enhancing technologies;
- information security;
- computer crimes;
- on-line identification;
- biometric identification;
- legal issues; and
- societal implications.

Other Computing Courses

Limited course time and content of different computing courses impose significant restrictions in introducing ethics in the material taught. However, it is always possible to find the right moment to attract students' attention in the desired direction. Students should become familiar with the concept of computer ethics defined as "the analysis of the nature and social impact of computer technology and the corresponding formulation and justification" [20]. Computing courses which incorporate ethical issues and have been thought by the author are as follows: *Introduction to Computer Applications, Computer Science I* and *II, Information Systems, Systems Analysis and Design, Database Systems, Database Systems Development and Applications, and Readings and Projects in Computing.*

Over the last few years, there has been a wonderful trend in publishing—an increase in the offering of computing of textbooks that contain ethics-related chapters in addition to specific computing topics. The courses mentioned above offer limited opportunities to teach ethical concepts and principles because of the wide range of topics that have to be covered in the given period of time. Today, most publishing companies offer a variety of textbooks which are appropriate for use in different disciplines, which include chapters discussing social and ethical issues, and which are related to multilateral computer aspects [1, 7, 11, 14-15, 23-24, 26].

Very often, the nature of the material taught in computing courses allows the instructor to shift the focus to ethical and moral issues related to software development. When students start designing and running their own computer programs, they are advised to consider whether the source code is understandable to other users. Students also need to think about the social impact of their software products. Initiating discussions about copyright concepts is also appropriate. In addition, students should acquire skills to protect their programs by implementing security techniques related to the respective hardware and software environments they are using.

Considering ethical issues is also appropriate in the *Information Systems* course. The textbook adopted for this course [15] offers a model for thinking about ethical, social, and political issues related to ISs by discussing the following moral dimensions: information rights and obligations, property rights, accountability and control, system quality, and quality of life. Students can learn about the basic rules that guide the decision-making process: Immanuel Kant's Categorical Imperative, Descartes' rule of change, the Utilitarian Principle, the Risk Aversion Principle, and of course, the ethical "no free lunch" rule. A valuable advantage of this textbook is the presence of multiple case studies that provoke detailed analysis and critical thinking. At the end of this course, students know the answers to the following questions [15]:

- *What ethical, social, and political issues are raised by the ISs?*
- *Are there specific principles for conduct that can be used to guide decisions about ethical dilemmas?*
- *Why does contemporary ISs technology pose challenges to the protection of individual privacy and intellectual property?*
- *How have ISs affected everyday life?*
- *How can organizations develop corporate policies for ethical conduct?*

In the rest of the courses, similar ways of incorporating ethical issues are used. They are always related to possible solutions for creating an efficient and safe computing environment through discussion topics such as computer crime, spyware, encryption, passwords, public and private keys, development of different types of ethical codes, backup and recovery procedures, firewalls, data and network security, and others.

Computer Ethics as an Out-of-Class Experience

Sometimes, the sheer number of assignments given to students can create indifference towards the course material. Breaking the stereotypical dynamics of the class room environment and generating appealing ideas related to the real world can lead to much better results. Basic out-of-class forms in teaching ethics that have been used by the author are the following:

- development of a variety of computing projects;
- organization of an undergraduate research-oriented seminar on ISs;
- design and maintenance of an information bulletin board for computing students;
- organization of quiz bowls;
- field trips to companies that are leaders in the introduction of IT in their business;
- involvement of students in organizing academic forums and activities.

Project Development

Using projects in computing-oriented courses is a well-known idea and an everyday practice. Their development requires the student to employ imagination, originality, considerable knowledge and skills, and, in some cases, multidisciplinary vision. The multi-year experience from the development of projects related to ethical and moral issues is presented below in a separate section of this paper.

Undergraduate Research-Oriented Seminar on Information Systems

Traditionally, undergraduate research presents an appropriate way to involve students in scientific or practice-oriented activities and to cultivate a taste for discovering new worlds in science and its applications. A student seminar on ISs was founded in the spring semester of 2002 by the author of this paper and continues to be held periodically for students with deeper interests and the desire to research cutting-edge computing. Usually, the presenters are students and, on rare occasions, invited faculty or IT professionals. Each presenter researches a selected topic in advance and shares findings and perspectives with other students by one of the following means: an electronic presentation, a demonstration of abilities to utilize given commercial product that is not part of the requirements for any course, or a talk and discussion with the audience. Some of the presenters were drawn to research topics that belong to the area of computer ethics, such as:

- database systems security;
- Internet piracy and legal aspects of the copyright;
- plagiarism in assignment development;
- history of hacking;
- fighting computer viruses;
- problems raised by the virtual reality applications;
- ISs related to ecological issues;

- biometrical identification;
- smart home technologies.

Students developing senior projects, required for the *Readings and Projects in Computing* course, are able to share their preliminary results and obtain valuable suggestions, advice, and overall help when presenting their work in front of the seminar audience.

This student-directed seminar explores ethical issues in the practice of working with computers. The goal consists of allowing students to gain better understanding of the respective categories. Instructors must prepare their students for the complex ethical challenges they are likely to encounter as future computing professionals and help them develop skills to work through ethical dilemmas that might arise in their daily work.

Design and Maintenance of an Information Bulletin Board

Students can choose from a variety of different service-oriented activities. One of them is the development of an information bulletin board. Along with materials that present specific computing-related issues, the board can display intriguing information about copyright, software and copyrighted music downloading restrictions and regulations, identity theft, existing codes of ethical conduct, information about the seminar on ISs, abstracts of senior projects, and many others. There is a special glossary section that is updated periodically and introduces students to cyberethics notions. The terms selected for an issue of the information bulletin board are usually logically connected. Such a group of ideas could be representative of the family of computer bugs, the diversity of computer hackers, the opportunities for online banking with related payment and safety options, common practices of backup and recovery, different methods for data and network security, or the set of law-related terms referring to intellectual products.

Quiz Bowls

One of the most exciting forms of introduction of ethics concepts and consideration of cyberethics applications and examples from real life is the quiz bowl. As already mentioned, students prefer such non-traditional forms of teaching. They like competitions that challenge their knowledge and skills and are delighted to be a part of a winning team, gaining the respect of their classmates.

Field Trips

Field trips to outside companies and organizations are enjoyable for students, as well. In 2003, Computerworld ranked Hershey Company as the top company in the U.S. that offers outstanding conditions for IT workers and the #7 U.S. Company in introducing IT to its business. In the fall semester of 2003, students majoring in computing programs visited Hershey's Data Center. The tour of the computer center was exciting; however, the follow-up discussion with the Data Center managers, which was scheduled for forty minutes but lasted two hours, proved how useful such activities could be. Questions related to professional relationships between managers and employees, hackers' attempts at the company websites, protection of data resources, backup and recovery policies, machine and human protection in emergent situations, obtaining of patents and trademarks by the company, and internships and career opportunities for students were all part of the discussion. The visit made a lasting impression on these students. They were able to hear directly from future employers what expectations of computing professionals were, including the ethical side of their human behavior.

Students' Participation in the Organization of Academic Forums and Activities

In September 2004, a conference on *IT in Education* which had a significant international participation was held at Elizabethtown College. The conference was a multidisciplinary forum to facilitate the exchange of information on current theory, research, development, and practice of IT in education and training. The goal of the forum was to stimulate the growth of ideas and practical solutions in the educational process in all academic disciplines. A guiding idea of the keynote speeches and some of the conference papers focused on teaching values while teaching computing courses. It was stated that this is important because all human activities require knowledge and skills in using computers by qualified and experienced people able to meet the challenges of the future to live in an ethical and highly developed society.

As the main organizer of the conference [35], the author was able to involve her students in a variety of ways. They participated in the development and the maintenance of the conference website and the IS that supported the overall conference information management and processing. Students were exposed to something vastly different from everyday class activities. For them, this was another way to understand the requirements relevant to computer ethics and specific for such events: privacy of information, confidentiality, copyright-related issues, correctness, data consistency, teamwork, etc. They were invited to share their research and practice-oriented results in the IT area and to be co-organizers of one of the main conference sessions, *Ethical Issues in the Computerized Classroom* [4, 35], including subtopics such as ethical agents in the information space, ethics in an information society, moral dimensions of IT applications, and others. Two of students had their research results published in the volume of the conference proceedings after a double-blind review. All students had free access to the paper sessions. They enjoyed the rare opportunity of meeting with scholars from all over the world and of hearing ideas and experiences from the area of their professional interests.

Especially interesting was their contact with the keynote speakers. One of them was Gerald Engel, Leonhardt Professor of Computer Science and Engineering from the University of Connecticut and President of IEEE Computer Society in 2005. He is well known for his computer ethics research. His talk about "The Range of the IT Field and Where We All Fit In" [35] was interesting to all conference participants. All computing, physics, and engineering students and faculty had the opportunity to meet with Professor Engel during a special event that was organized separately from the conference. He delivered a presentation on *How to Become a Good IT Professional* and provoked spirited discussions, which also included the IEEE Code of Ethics. Another keynote speaker, Paul Tymann, professor of computer science at the Rochester Institute of Technology, offered an interesting presentation on "Bioinformatics: A Recipe for Inter-Disciplinary Collaboration" [35] which also considered ethical aspects. The third keynote speaker, Dennis Christopher, senior scientist at the NASA Goddard Space Flight Center spoke about "Navigating NASA: How to Find Information in a Labyrinth of 2.5+ Million Web Pages" [35]. Surfing the Internet is always related to safety issues and the students had the rare opportunity to learn about concrete practices applied by NASA for online access to information resources and network security. Lastly, the curator of the ENIAC Museum at the University of Pennsylvania, Paul Shaffer, was the invited speaker of the panel session on computing history. His presentation titled "ENIAC: Dawn of the Age of Information" [35] triggered questions around one of the most famous ethical controversies known to international computing society—the birth of the first digital computer.

The experiences from this conference were useful to students. They were able to actively participate in its organization and hear the opinions of established researches and scholars on different topics, thus gaining valuable personal and professional experience.

Variety of Projects

Teaching computing sciences involves a significant number of projects periodically assigned to students. Developing projects is a multilateral process having different aspects (figure 1). It is an opportunity to reinforce the academic requirements relevant to a given course. Students majoring in specific computing areas such as Computer Science, Information Systems, Computer Engineering, and others are able to obtain knowledge and skills in using a variety of commercial software products and a strong background in practical professional activities. In this way, they prepare for their future jobs and for pursuing higher academic degrees. These educational challenges lead to specific visions expressed by:

- students;
- academic institution;
- industry; and
- society.

Students' goals gravitate towards acquiring professional skills that allow them to use the most recent software products. This helps students be immediately involved in internship activities or other current or future job responsibilities, in addition to their academic studies.

Academic institutions focus on covering curricula requirements, providing a strong theoretical background to students, incorporating practical orientation of the covered material, and teaching appropriate ethical and moral principles.

Industry needs knowledgeable workers, skilled IT users, and software developers who are capable of accomplishing various computing activities.

Society's interest lies in having valuable members capable of making the right choices about their ethical behavior and acting according to pre-established moral principles.

Figure 1
Aspects of Developing a Computing Project

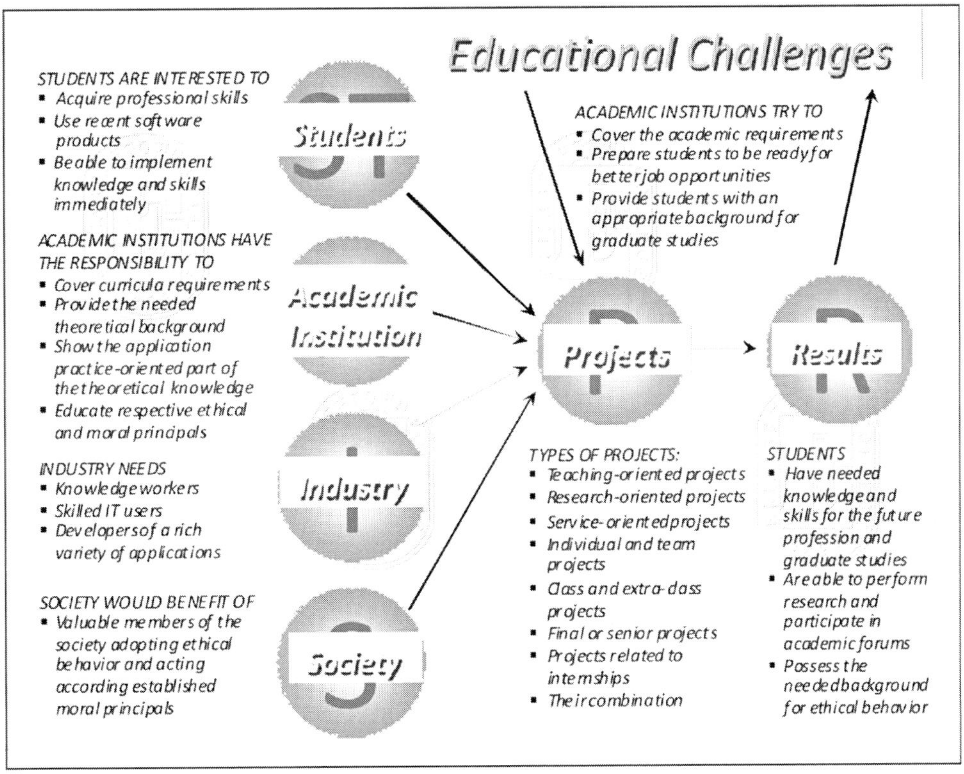

There exist different project types. According to their orientation they could be teaching/learning-oriented projects, research-oriented projects, service-oriented projects, or a combination. Project collaboration is relevant to two other project types: individual projects and team projects. Projects could be developed as a class assignment or as an out-of-class activity. The most valuable group, however, consists of projects which involve service-learning features because of the increase in student motivation as well as the related moral rewards.

The development of computing projects reflects the above four views and leads to the achievement of important results such as providing opportunities for undergraduate research and practice-oriented applications, acquiring relevant knowledge and skills needed for the future students' profession and possible graduate studies, and adopting appropriate professional ethical behavior.

The variety of computing projects developed in a liberal arts college environment that are related to teaching ethical and moral values is described below. Concrete examples have been considered in different courses taught by the author. All projects which were developed by students aimed to combine the practical application of IT knowledge and skills with ethical aspects relevant to the college community, to external companies and organizations, or to society and families in particular. Students collaborate with different people and try to find a solution that would be right not only for them. Sometimes they face dilemmas and have to make decisions that involve the assessment of values in particular cases [29]. That gives them the opportunity to consider different scenarios as possible outcomes, to perform analysis, and to make personal choices. As a result of the project's development, students acquire a higher level of confidence in their abilities as computing professionals and get to experience the fruits of their decisions and actions.

Several basic criteria are applicable when differentiating the possible types of projects. They are related to teaching and research factors, collaboration aspects, in-class or out-of-class work, internal or external to the college activities, service learning involvement, and others. Usually, the projects incorporate features from more than one project type and that is what makes them even more valuable. Different categories of projects are presented below.

Project Orientation Factor

Teaching/Learning-Oriented Projects. The first type of project is closely related to the material taught in a discipline and is part of the course syllabus in the respective major. Usually, there is no special course on computer ethics included in a computing curriculum. However, the content of other computing courses offers the opportunity to discuss ethical and moral issues. In most the cases, newly published textbooks contain chapters discussing such topics. Often, students who attend these courses must develop projects individually or as participants of a team. They are required to write essays or even a sample code of ethics, to analyze case studies, to present results electronically, to create final solutions such as programs, drawings, diagrams, ISs, and websites, and also to use specific software products, integrated packages, systems, and computer-aided software engineering (CASE) tools. Such projects have been broadly explored by instructors [10, 18, 32]. Feedback from alumni testifies to the usefulness of such projects. For example, the most appreciated database projects are related to the SQL query design and processing and the development of database-driven websites. This could be explained by the recent interest of companies in e-commerce and m-commerce applications. The environment of such projects requires interaction with users and imposes the necessity of considering ethical issues.

Research-Oriented Projects. The second type of projects accentuates undergraduate research. Students developing senior projects are interested in obtaining an official proof of their knowledge and professional skills that can be pointed to or cited during job interviews or on graduate schools applications. The author organizes a seminar in ISs inviting students to share their knowledge in current computing topics and to present results obtained in their research projects. Usually, the seminar is attended by students who show a great interest in different areas of computing and are looking to find an academic forum to present, discuss, and defend their ideas. In order to be best prepared for the public presentation of their results, students' cross-check multiple sources of information and read impressive amounts of materials related to the respective topic. The role of the instructor is also very important. Offering advice and suggesting useful books and journals is critical. By working on such projects, students come to realize what their real interests are. They acquire a taste for this type of work, especially if they succeed in presenting their research results at professional workshops or conferences. Sometimes, the preparation of such projects is associated with the decision made by students to enroll in a graduate academic program. Attending other student presentations offers the opportunity not only to learn about new research areas but also to establish useful multidisciplinary relationships and to generate new ideas. Again in this case, when developing research projects, students are advised to take into consideration their responsibilities as computing professionals and to act according to the norms of ethical conduct similar to those listed in the Code of Ethics for Professional Conduct of ACM or the IEEE Code of Ethics. Both organizations are international leaders in the area of computing and are among the first organizations having developed such codes.

Service-Oriented Projects. In the liberal arts college environment, the most rewarding projects are those related to service activities which are the duty of the students and the faculty as members of the college community and society. Because of the importance of this group of projects, a special section has been dedicated to them. It offers an overview of several projects which have been developed and implemented by students who also encountered ethical issues in the creation process.

Project Collaboration Factor

The practice of teaching introduces different forms of collaboration:

- collaboration among instructors and students;
- collaboration among students;
- collaboration among students and external companies and/or organizations.

Usually, the last group of projects is related to students' internships. All approaches in the project development could be applied [6, 30].

Individual Projects. Most students, especially those who have excellent performance in the subject taught, prefer individual projects. Senior projects are typically individual.

Team Projects. The individualistic mode of social behavior found in certain nations calls for the use of team projects to foster an appreciation of group work, especially in the area of computing. The valuable group experience gained from such projects has been long recognized. College instructors also use this type of project to teach principles related to computer ethics. Usually, they apply two types of grading for such a project: grading the individual student's results and grading the group results. Students who need help could benefit from working this way and a better overall performance of class work could be achieved.

Internship-Related Projects. Prior to or during the junior year of study, nearly all students like to have an internship directly related to their future profession. They hope that it will progress into a job offer. Even if an offer is not obtained, they can gain valuable professional experience and learn how to act ethically in a real working environment. They try to do their best in developing and implementing different valuable projects. Such projects are highly encouraged and it is suggested that they combine them or parts of them with the final projects developed for the *Readings and Projects in Computing* upper level course. This way, students continue to prepare for a future job that corresponds to their individual professional interests.

Projects as Class Activities or Out-of-Class Activities

There are two additional approaches to project development: projects prepared as an in-class activity and projects assigned as homework without direct oversight from the instructor. They could be considered as a part of the sections already discussed above which focused on cyberethics as a class and out-of-class activity.

Projects in the Classroom. Students are interested in acquiring practical knowledge when using different commercial software products. Such knowledge can be gained by working on specific projects. The "learning by example" and problem-solving methods are quite useful and students enjoy such classes in the computer lab. They can easily grasp the corresponding knowledge and skills in a more informal atmosphere. This is also the time when the instructor can assess student performance and extend personalized help to them.

Projects Outside the Classroom. The real proof of knowledge and skills can be demonstrated by developing multiple project assignments. Some of them can relate to a specific topic presented by the instructor in class. Sometimes, students are asked to develop comprehensive projects, to draw on knowledge from different computing areas or even different disciplines. The projects contain substantial research elements. These projects are valuable but time-consuming and this is the reason why they must be prepared outside the classroom.

Faculty-Oriented Projects

Elizabethtown College offers grants to its faculty for developing different types of projects. These projects represent an interesting and wonderful opportunity to involve students as participants. In this way, students can gain useful experience in solving real problems, working in teams, assuming responsibilities, and also learn how to develop products that represent intellectual property. They are involved in the process of defining problems, generating ideas for resolution, analyzing these ideas, and making a well-reasoned decision.

Faculty Grants. The author was awarded a faculty grant for the development of a lab manual for her *Information Systems* course. The help of students in creating appropriate case studies, examples, exercises, and project assignments was very useful and appreciated. The satisfaction from the end result, obtained through joint effort, respectful interaction and collaboration, and valuable work experience was rewarding.

Strategic Grants. According to [22], the goal of a project that is related to a strategic grant is "to offer innovation for a department, program, or other specific advantage for the college, including academic, research, and entrepreneurial projects; and/or to lead to an external grant application." Such a grant was awarded to the author for organizing a conference on *Information Technology in Education* [4]. This activity also included several student projects such as the development of the conference website, creation and maintenance of the corresponding IS, and overall assistance in accomplishing conference tasks. The significance of the professional skills gained by students was not the only positive outcome. Students learned about the process of reviewing, accepting, publishing and presenting conference papers. They also understood how to officially recognize and properly cite the work of others without violating copyright norms. They obtained valuable advice from distinguished scholars who also taught them how to act ethically as computing professionals.

Mixed Projects

Most projects include features from different project types. There is a wide range of possibilities to combine teaching, research and service-oriented activities [25]. This is so valuable because it incorporates the diversity of aspects related to all respective projects in consideration. Students are even encouraged to work on interdisciplinary projects that would allow them to gain broad and useful experience [34]. The combination of the student's major with another appropriate major or minor occurs frequently at liberal art colleges. This offers a prime opportunity for the development of interesting projects that reflect the student's vision, creativity, and strong interest related to multiple scientific and practice-oriented topics.

Service Learning Projects

According to [34], service learning represents "a method under which students learn and develop through active participation in thoughtfully organized service experiences that meet actual community needs, and which are coordinated with a formal educational institution to address and support an academic curriculum." Developing projects relevant to the concept of service learning is challenging. However, successful ones yield high rewards. Generating an appropriate theme for a specific service-learning project is especially important. By developing such projects, students prove their ability to work with widely used software products such as database management systems (Microsoft SQL Server, Oracle, MySQL, Microsoft Access) and specialized software for development of database-driven websites (ASP, PHP, XML, Adobe ColdFusion, Adobe Dreamweaver, VBA, and others). One of the most precious aspects of working on such computing projects consists of the necessity of resolving situations related to ethical and moral issues and the required decision making and assessment of corresponding values when choosing concrete technological solutions. Students go through various phases during their projects:

- selecting the right legal software to be used;
- implementing the results of others in their work by properly citing the respective references;
- taking into account privacy problems when using the final product they have created;
- establishing respectful professional relationships with people involved in the project activities;

- thinking about the social impact of the results of their efforts and time spent for the project;
- choosing the most appropriate human-computer interface for the software product they have created;
- learning from the experience of others contained in codes of ethical conduct for professionals and users;
- serving by doing something useful for others; and
- feeling satisfaction from their work and the deserved appreciation of the community.

College Community Related Projects

The author, serving on different college committees and councils, initiated the development of projects focused on actual community needs: a survey linked to alcohol usage and driving on campus accompanied by appropriate data analysis and visualization; an on-line survey related to music piracy and copyright issues; projects inspired by field trips to leading computing-related companies; development of a website and a supporting IS for a conference organized by the college, and others. Such projects attract the recognition of the college community, and this fact motivates the students to develop new and better projects. Sample projects developed under the academic advising of the author are presented below.

Disability Information and Elizabethtown College. Being a member of the Campus Life Council, the instructor knew that an IS and a corresponding website, providing extensive information on campus accommodations for disabled people, would be extremely beneficial. A series of projects spanning different classes was created to fulfill this need. It was important to motivate all students to do their best in this endeavor. A team project considering aspects of systems design and documentation was assigned to students attending the *Systems Analysis and Design* course. Students checked all of the buildings on campus to determine their accessibility. They created detailed floor plans of the buildings using appropriate software CASE tools to provide visual information about entrances, elevators, area parking, academic departments, and offices in each building (figure 2). In addition, they estimated the corresponding times and distances between buildings. Parts of the project were accomplished in class, but most of the work was done outside the classroom and later in the computer lab.

The results from the team project were used in the development of two other comprehensive student projects assigned in another course, *Reading and Projects in Computing*, as individual projects. The first of them resulted in a website offering on-line information for the disabled about campus buildings, facilities, and parking areas under development. The second project created a corresponding IS containing the needed information and allowing its rapid processing.

Figure 2
Pictures and Floor Plans Indicating the Accessibility for the Disabled

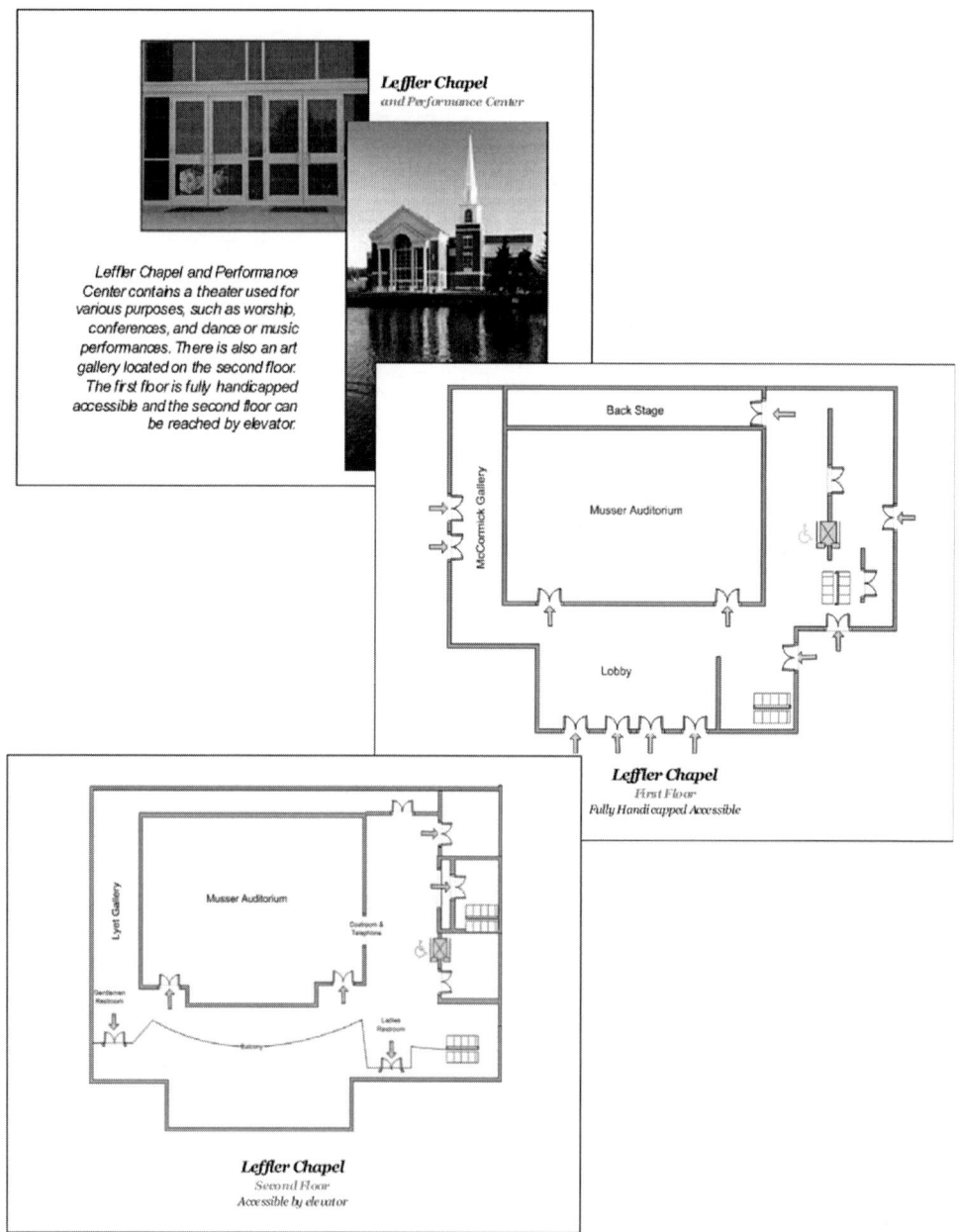

Ongoing construction and renovation activity on campus had an influence on these projects over the next few years. Students attending the respective classes during the next years updated the projects by introducing information about the new buildings, parking areas, and changes in the existing ones.

The impact of these joint projects was measured by their role on campus. They were recognized by the Campus Life Council. The results obtained were used in the virtual tour of the college campus for the new college website. The data gathered was also found to be a very useful source of information by the Landscaping Committee in the development of official campus maps. Disabled people living on campus or visiting the college can also obtain a booklet containing the needed information about campus accessibility. The projects also offered relevant opportunities to consider ethical issues with regards to cyberethics and to discuss them with the students.

On-Campus Usage of Alcohol and Driving Survey. This project was also brought to the attention of the Campus Life Council. A student majoring in Business Administration proved his computing knowledge and skills by processing information related to the college campus and community. The student performed statistical analysis and displayed the results through multiple data visualization techniques producing informative charts and diagrams. The statistics gathered through this project were analyzed by different college divisions and useful conclusions were drawn. Discussing the respective results with students in class generated intense discussions and helped them make choices about personal behavior, responsibility, and safety on campus.

Music Piracy and Copyright On-Line Survey. The traditional concept of academic integrity has expanded its area of coverage. Now, it includes non-traditional aspects related to cyberethics because of the influence of IT on overall academic activity. A blend between the goals of the academic integrity committee objectives and a senior project led to the development of a database-driven website. This website was used to publish an anonymous survey in the form of a set of multiple choice and true/false questions about music piracy and copyright issues related to Internet sources. All students were invited to complete the survey. The purpose of the project was to offer useful information to students in an interesting and non-traditional form. The most important result was not the information gathered but the fact that the students were exposed to the legal aspects of copying music and documents from the Internet. They also learned about the existence of organizations authorized to control these activities.

Information System and Online Application Form for Committee Activities. Service activities are especially appreciated at Elizabethtown College and exemplify the motto of the college—"Educate for Service." Faculty and staff participate in different committees and councils that organize and control various academic and campus-related activities. A student developed an IS and a database-driven website as a senior project to help the Nominating Committee maintain complete information about all committees and councils: current participation, announcement of vacant positions, collection of proposals for participation received through completing an online application form, archiving the information, and others.

Online Form for Faculty Grant Reports. Another committee-related project was developed as a student senior project. It consisted of creating a website that provides an online form. This form could be used to generate a report about the activity accomplished through a faculty grant awarded by the Professional Development Committee.

Career Services: Internship Database. After the sophomore and/or junior years of study, students are interested in having an internship that allows them to implement their professional knowledge and skills in real-life situations. This is a serious step for them because in many cases, internships turn into job opportunities and the selection of the company or organization is important. The Career Services college division organizes different activities on campus and maintains useful online information that can help students in making their choices. In the spring of 2005, an external grant was awarded to this division by the state of Pennsylvania through the Lancaster County Workforce Investment Board. The grant was related to assisting students with workforce development skills and encouraging internships and employment with the county and the state of Pennsylvania. One of the accomplishments under this grant was achieved by a senior student, who developed a database and a corresponding website to help students find current internship offerings. The website also included functionality for employers to post new internships and related job openings.

Dining Services Interactive Database. An experimental IS and a corresponding website were developed by a senior student as his final project to allow students to estimate the daily quantity of food, measured in calories, offered in the college dining room. The IS also included the product description and information about the respective content of fat, saturated fat, fiber, sodium, cholesterol, protein, and carbohydrates. The close collaboration between the student and the Director of the Dining Services resulted in a successful and useful project. Students who volunteered to experiment with this IS provided positive feedback.

Elizabethtown College Used Book Website. A very appealing idea to students, consisting of advertising information about used textbooks, generated the creation of a website and the underlying database. Students were able not only to sell their old textbooks, but also to upload what they would like to buy. The website offered user-friendly interaction tools for the fast exchange of information.

Virtual Ad Board. This project represents a generalization of the previous project by offering the opportunity for students to buy and sell not only used textbooks, but also computers and accessories, audio and video equipment, cars, furniture, and many other products. The respective software and website were intended to be distributed through the local college network.

Media Website. Students are involved in a rich variety of out-of-class activities on campus. One type of activity is related to collecting, creating and distributing information related to the college community by using different media opportunities. Students, trusting also the experience of faculty and staff, publish their *E-Townian* newspaper, maintain a radio channel, and broadcast college-related information on their own TV channel. News about current and upcoming events disseminated through all these three types of media were included and maintained on a website that was developed by a student as a final computing project.

Elizabethtown College ITS Knowledge Database. As technology advances, the average user must strive to keep abreast of current trends. With the advent of the Internet and increased complexity of operating systems, users encounter more problems than ever before. Given the wide range of computer viruses and spyware as well as thousands of different hardware configurations, errors are bound to happen, especially on a complex college network. Knowledge bases are becoming increasingly more common as end users try to solve their computer issues on their own. A senior student created a knowledge base as a final project. It was oriented towards network users who encounter computer problems. The college network is maintained by the Information Technology Service (ITS) department. Users are able to solve different problems by themselves by using the knowledge base. This is especially important during the busiest time for the ITS, which occurs at the beginning of each semester. The role of the knowledge base is to alleviate the heavy call volume at the Help Desk. In addition, the system represents a valuable training tool for new Help Desk technicians.

Projects Related to External Companies and Organizations

The academic background, enthusiasm, creativity, and rich imagination of students find outlets through the development of projects related not only to the college community, but also to external companies, organizations, and other social institutions. Such projects have been strongly encouraged because they offer the opportunity of gain useful experience from applying students' knowledge and skills in solving real problems that require a professional background and corresponding ethical behavior, as well. Examples of such projects created as internship solutions and/or individual senior projects are presented below.

Electronic Diaries in Clinical Trials. Security has become a major area of concern due to the growth of networks and the increasing number of electronic devices that connect through them. This is especially important in the medical industry, where protecting the privacy and security of patient information is of the highest concern. In 1996, Congress passed the Health Insurance Portability and Accountability Act (HIPAA), which established rules for handling and securing medical records. Organizations also need to take into account mobile devices such as notebooks, tablet personal computers (PCs), and personal digital assistants (PDAs) that may be accessed to store, process, and visualize sensitive information in order to comply with respective HIPAA standards.

An internship project discussed the use of electronic diaries in clinical trials. This particular electronic diary was used by the Childhood Asthma Research and Education Network (CARE) in a Montelukast or Azithromycin for Reduction of Inhaled Corticosteroids in Childhood Asthma (MARS) clinical trial pilot study at Hershey Medical Center in Pennsylvania. Asthma is the most common chronic childhood disease in the U.S. and the CARE Network is conducting several studies for children with asthma and sharing their findings with the health care community to help understand and treat this disease.

There were a significant number of subjects who required the usage of a PDA in this study. Each subject used the device to fill out diary information on daily basis. The PDA had a program that sent data via a 56k modem to a server located at the Hershey Medical Center. For this project, the student used AppForge Crossfire to write a Microsoft Visual Basic.NET program on a Palm Tungsten E2 handheld. The clinical data was sent to an Oracle database and could be accessed by centers through a database-driven Web application using Macromedia ColdFusion. The project was successfully implemented at the Hershey Medical Center and currently it continues to be used by thousands of patients.

Information Systems in Hospital Operating Rooms: Using Radio Frequency Identification to Safeguard Patients and Their Medical Records. At a time when healthcare costs are soaring due in part to patients being harmed by avoidable errors, such as harmful prescription drug interactions, Radio Frequency Identification (RFID) is seen by many as a way to combat these errors. The problem causing these errors often relates to patient records not being stored electronically in an IS, which could detect a variety of potential harmful drug interactions, as well as other dangers. The other potential cost-saving benefits of RFID for hospitals include asset tracking and monitoring, inventory control, specimen identification, and transfusion safety. In researching RFID, a senior student developed a final project to explore its potential benefits in saving lives, as well as its potential dangers in divulging patient medical records. It is clear that hospitals using RFID technology, along with RFID related ISs and wireless networks need to take security precautions seriously. The project included a business case for RFID technology in the hospital. The project's conclusion stated that RFID return on investment can be significant if it is implemented in a responsible manner. Also the project included a sample IS, which demonstrated a simple implementation of RFID technology and its potential benefits. Overall,

RFID is clearly the future of better hospital care, once it becomes fully mature, standardized, secure, and cost effective.

CompuDent: Software for Dental Professionals. Dental offices need easy-to-use, full-featured dental software that integrates dental practice management and also simplifies the billing and scheduling for patients. Such an experimental IS was created by a senior student who had an internship at a dental clinic and was involved in the processing and management of the working data. Information about patients and their health is very personal and therefore, the patients' privacy needed to be protected as required by law. ISs aspects related to data security were also considered in this project.

The C2K Agreement Website. Organizations rely heavily on the Internet because it has emerged as an effective mainstream tool. It allows them to communicate electronically and to share knowledge and vital information, which in turn reduces staff work load and paperwork. Because of advances in Internet security, businesses and organizations can make greater use of the Web without as much concern for safety and integrity of their websites. Not only can organizations provide useful information on the Web but they can also receive feedback and performance evaluations from the public and fellow corporations.

The Maryland Department of Natural Resources (DNR) is using the Internet to present useful information about the DNR that the general public can access easily from any computer. Providing this service benefits the DNR and the public. The public can get the information they need more quickly and the DNR benefits by keeping the public well informed of news and developments within the natural resources field. DNR's mission clearly states that they intend to "preserve, protect, enhance and restore Maryland's natural resources for the wise use and enjoyment of all citizens" [37] and their website affords them the opportunity to post vital up-to-date information on how they are achieving these goals.

A graduating student developed an internship-oriented final project by taking advantage of her knowledge about the Internet. She created a website for the DNR that utilized a database to communicate facts and information about how the State of Maryland is fulfilling the department's goal of protecting and restoring the Chesapeake Bay. This website was developed to track progress in meeting the specific commitments of the Chesapeake Bay 2000 Agreement (C2K). The website provided the community with needed information that aided in understanding the progress on each of the 103 commitments contained in the C2K Agreement.

Practical Application of Information Technology: A Field Studies Database. The development of interdisciplinary undergraduate research projects appeals to students. An example of such a senior project was developed by a student who majored in both Information Systems and Biology.

The focus of the project was to develop and implement an IS for a new biological company that provides environmental consulting for small and large building projects. This service includes wetland delineation as well as presence/absence surveys for endangered and protected species. The company is one of only a handful of groups that are qualified to provide these services. During their field studies, they keep records of nearly fifty different species from categories including turtles, snakes, frogs, salamanders, and lizards. Their studies often include up to six visits to each site with the possibility of several animals per site. It is easy to imagine that with multiple sites, the data could accumulate quickly.

The IS was designed to be user-friendly and to allow the user to insert, delete, and update records, as well as view data with or without setting result criteria. The system made it possible for the company to gather large amounts of information on the field and then enter that data into the database in an organized manner, thus allowing for quick and easy access to information regarding different sites, as well as allowing predictions about other sites to be made with minimal field research.

Electronic Management of Foreign Currency. Another internship-related senior project consisted of the development of an IS that manages foreign currency operations at a Pennsylvanian bank. After the implementation of the system at the bank, the student who created it was hired as a consultant to introduce additional business functions, some of which are still in use after five years.

Student Tracker. Establishing good relationships with parents is important for elementary, middle, and high schools. The learning process can be improved if parents have full information about the class performance of their children. A student project considered the development of a database-driven website that offered a close interaction between teachers and parents, information related to different school events, programs, out-of-class activities, and financial issues, as well.

Rental of Equipment for Winter Sports. A student who worked for a company that rents equipment for winter sports choose to develop an IS that organized and managed useful inventory and operational data for the company, namely equipment description, dates, names of customers, and many other data organized as database tables. The IS could also be used to analyze customer preferences and orient managers on how to make future investments by purchasing new equipment.

Roman Catholic Community Web Service Information Portal. In choosing a senior project, a student decided to construct a Web portal that included information about a local Roman Catholic Community Church with mass times, pastors, bulletins, maps, and school programs. Many hours of research were dedicated to finding accurate histories of all the parishes and to selecting the right software for the project development. Users really appreciated the final result after the system implementation, though.

Shopping Cart Information System. A family-owned business in central Pennsylvania was looking to re-design its website for its volleyball retail store in order to reach a broader customer base. Although there were numerous off-the-shelf solutions, none of them provided the needed flexibility at a reasonable price. This was the reason for creating a new IS as a final project by a student whose father was the owner and manager of the company.

There were a number of requirements that were imposed on the IS by the business in regards to how the system needed to be operated. Since the store also operated a retail location, it was important that the IS be easily accessible. The employees needed to have access to change current stock levels and view in real time whether products were in stock or needed to be back ordered. Obviously, the employees also needed access to order records that allowed them to process and ship each order as it arrived.

The second major requirement of the IS was that new products had to be easy to add. The inventory at the store changed at a very high rate following supply and demand. The inventory also changed regularly, as the vendors that the store used released new lines of clothing, shoes, and equipment on a regular basis. Therefore the website had to be updated with the new products while the old ones were removed.

A third major component that was required in the new website was linked to the underlying database and included additional information not related to the e-commerce system. The company also operated the Lancaster Area Volleyball Association (LAVA) and numerous tournaments for the Amateur Athletic Union (AAU). Thus, the company wanted to make much of the information about these two organizations available on their website. They also wanted to utilize these two sponsored programs to help generate traffic to the retail products they had available online.

Conclusions

Instructors teaching computing courses should foster a creative academic environment that allows their students to develop useful moral reasoning skills. They should also provide them with a strong foundation for identifying and responding to ethical issues in their future professional life. The Australian Department of Education, Science and Training has developed a list of values [21] that could be framework when teaching IT at academic institutions. The list includes:

- care and compassion;
- doing your best;
- fair go;
- freedom;
- honesty and trustworthiness:
- integrity;
- respect;

- responsibility;
- understanding, tolerance, and inclusion.

All examples of in- and out-of-class activity described in the paper and related to teaching values in computing courses at a liberal arts college through theory and practice, incorporate values among those listed above. Obtaining theoretical knowledge and practical skills in developing and using software products and possessing a strong background in professional activities is important for students. Understanding the values relevant to technology would allow them to rationally and thoughtfully evaluate all of the considerations that define the best solutions in their professional daily work that would provide the highest benefit to society. With these knowledge and skills students will be prepared for the highly competitive real world, outside of their academic institution. They will be ready to become valuable professionals in their area of specialized study and members of society who possess a good understanding of a broad variety of academic disciplines and corresponding ethical behavior.

References

[1] Beekman, G. & B. Beekman. *Tomorrow's Technology and You*, Prentice Hall, 2008.

[2] *Computing Curricula 2001*, http://www.computer.org/education/cc2001. May 11, 2004.

[3] *Computing Curricula 2008*, http://www.acm.org/education/curricula/ComputerScience CurriculumUpdate2008.pdf. January 12, 2009.

[4] *Conference on Information Technology in Education*, Elizabethtown College, http://www.etown.edu/ITEconference2004. July 10, 2006.

[5] Edgar, S. L. *Morality and Machines,* 2/e, Jones and Bartlett, 2003.

[6] Finelli, C. *A Team-Oriented, Project-Based Freshmen Problem Solving Course: Benefits of Early Explore*. 29th ASEE/IEEE Frontiers in Education Conference, 1999.

[7] Fuller, F. & B. Larson, *Computers. Understanding Technology*, 3/e, EMC/Paradigm, 2008.

[8] Ghafarian, A. *Integrating ethical issues into the undergraduate computer science curriculum*. The Journal of Computing Sciences in Colleges, December 2002, Vol. 18, Issue 2, pp. 180-188.

[9] Greening, T., J. Kay & B. Kummerfeld. "Integrating Ethical Content into Computing Curricula," *Proceedings of the Sixth Australian Computing Education Conference,* Dunedin, NZ, 2004.

[10] Groth, D. & E. Robertson. *It's All About Process: Project-Oriented Teaching of Software Engineering*. 14th Conference on Software Engineering Education and Training (CSEE&T), Charlotte, NC, USA, 2001, pp. 7-17.

[11] Haag, S. & M. Cummings. *Management Information Systems for the Information Age*, 7/e, McGraw-Hill, 2008.

[12] Huff, C. & D. Martin. *Computing consequences: A framework for teaching ethical computing.* Communications of the ACM, December 1995, Vol. 38, Issue 12, p. 75.

[13] Kizza, J. M. *Ethical and Social Issues in the Information Age.* 3/e, Springer, 2007.

[14] Lauckner, K. & M. Lintner. *The Computer Continuum*, QUE, 1999.

[15] Laudon, K. & J. Laudon, *Management Information Systems: Managing the Digital Firm*, 10/e, Prentice Hall, 2007.

[16] Mayer Hanchey, C. *Yes, you can teach ethics!* The Journal of Computing Sciences in Colleges, March 2002, Vol. 17, Issue 4, pp. 145-153.

[17] McGuffee, J. W. *Tools for teaching computer ethics*. The Journal of Computing Sciences in Colleges, April 2002, Vol. 17, Issue 5, pp. 161.

[18] Mendoza, R. A. & H. J. C. Ellis. *Knowledge Transfer in IS Education: A Project-Based Pedagogical Approach to Bridging the Applicability Gap.* The Journal of Computing Sciences in Colleges, vol. 18, no. 3, February 2003, pp. 87-99.

[19] Meyenn, A. "A Proposed Methodology for the Teaching of Information Technology Ethics in Schools," *Selected Papers from the Second Australian Institute Conference on Computer Ethics*, Vol. I, November 2000, pp.67-72.

[20] Moore, G. E., *Principia Ethica.* http://books.google.com/books?id=2Of_A6PZskgC&dq=Principia+Ethica,+G.+E.+Moore&printsec=frontcover&source=bn&hl=en&sa=X&oi=book_result&resnum=4&ct=result#PPP1,M1. September 11, 2008.

[21] *National Framework for Values Education in Australian Schools.* Department of Education, Science and Training. Canberra: Commonwealth of Australia, 2005.

[22] *New Faculty Grants and Incentive Program*, Elizabethtown College, 2004.

[23] Pfaffenberger, B. *Computers in Your Future 2003*, Complete, Prentice Hall, 2003.

[24] Ricardo, C. *Databases Illuminated*, Jones and Bartlett, 2004.

[25] Sanderson, P. *Where is the computer science in service-learning?* The Journal of Computing Sciences in Colleges, October 2003, Vol. 19, Issue 1, pp. 83-89.

[26] Shelly, G., T. Cashman & H. Rosenblatt, *Systems Analysis and Design*, 7/e, Course Technology, 2007.

[27] Spinello, R. A. and H. Tavani, Editors. *Readings in CyberEthics*, 2/e, Jones and Bartlett, 2004.

[28] Spinello, R. A. *CyberEthics. Morality and Law in Cyberspace*, 3/e, Jones and Bartlett, 2006.

[29] Staehr, L. J. and B. Graeme. *Using the Defining Issues Test for Evaluating Computer Ethics Teaching.* IEEE Transactions on Education, May 2003, Vol. 46, Issue 2, p. 229.

[30] Stein, M. V. *Using Large Vs. Small Group Projects in Capstone and Software Engineering Courses.* The Journal of Computing Sciences in Colleges, vol. 17, no. 4, March 2002, pp. 1-6.

[31] Tavani, H. *Applying an Interdisciplinary Approach to Teaching Computer Ethics*. IEEE Technology & Society Magazine, 2002, Vol. 21, Issue 3, p. 32.

[32] Tuttle, S. M. *Practical Lessons from Experience with the Database Design Course Project.* The Journal of Computing Sciences in Colleges, vol. 18, no. 2, December 2002, pp. 32-42.

[33] Walbert, M. S. *Educating Illinois Values and the use of Technology in Teaching and Learning.* http://my.ilstu.edu/~mswalber/UTW_021.htm. January 7, 2009.

[34] *What is Service-Learning?* University of Colorado at Boulder, http://csf.colorado.edu/sl/what-is-sl.html. August12, 2007.

[35] Zlatarova, F., Editor. *Proceedings of the Conference on IT in Education*, September 18, 2004, Elizabethtown College, PA.

[36] http://www.acm.org/serving/ethics.html. September 23, 2008.

[37] http://www.dnr.state.md.us/mission.asp. January 16, 2009.

[38] http://www.ieee.org/web/membership/ethics/code_ethics.html. November 29, 2008.

Problems of Technology

Alan Kim

In the *Discourse on Inequality*, Rousseau contrasts natural and civilized man. The former has "all his forces constantly at [his] command, [is] always prepared for any eventuality, and [is] always, so to speak, altogether complete in [himself]."[1] The latter by contrast has delegated his powers to his tools and equipment, and thus compromised his integrity. If one were to strip civilized man of his things and set him against the savage, it would quickly become apparent who is stronger absolutely. Rousseau does not claim that a "natural man" or "savage" ever existed.[2] Nevertheless, the notion of a pre-technological stage of human development is important as a limit concept in the *Discourse*, as a foil for the always overly civilized condition of the modern. One can easily see how this limit concept is generated. Taking the present human condition as a starting point, subtract one technical advance after another, thus revealing an ever simpler and "purer" human life-world. Eventually, you will have returned to an Edenic state: human beings stand there, naked, healthy, and innocent of clothes and tools. The moment they reach down to pick up a hammer-stone, or pluck a fig leaf to cover their nakedness, must now appear as the first moment of a history of corruption and decline. Technology is a symbol of the Fall.

The same tendency to consider the human in abstraction from technology is evident among those who place their hopes for human progress in the constant development and dissemination of new technologies. The technophile, like the technophobe, considers technology itself to be a "tool" that can be picked up or dropped at will. On this view, too, the primal human is seen as naked, but his nakedness now is a sign not of strength and innocence, but of deprivation: his lack of technology signals his savagery. Civilization is seen as the advance by which "primitive" man sheds his feathers and loincloths and exchanges them for top hat, tails, and walking stick, as in George Catlin's double portrait of the Assiniboin chief, Wi-Jun-Jon, before and after his journey East.[3]

These familiar and commonly counter-posed perspectives on technology both commit the fallacy of *abstracting* the human from the technical, and so fundamentally distort our true relationship to the tools and equipment that sur-round us. The belief that the human being might be stripped of technology, yet remain essentially himself, until recently dominated even the scientific discourse on human evolution. The received view, as Sherwood Washburn puts it, "held that man evolved nearly to his present structural state and then discovered tools and the new ways of life that they made possible."[4] Yet as he argues in "Tools and Human Evolution," pre-human hominids already made and used tools: "It follows that the structure of modern man must be the result of the change in the terms of natural selection that came with the tool-using way of life."[5]

Washburn focuses on human bipedalism and dentition. Regarding the former, he writes,

> The fossil record thus substantiates the suggestion, first made by Charles Darwin, that tool use is both the cause and the effect of bipedal locomotion. Some very limited bipedalism left the hands sufficiently free from locomotor functions so that stones or sticks could be carried, played with and used. The advantage that these objects gave to their users led both to more bipedalism and to more efficient tool use. English lacks any neat expression for this sort of situation, forcing us to speak of cause and effect as if they were separated, whereas in natural selection cause and effect are interrelated.[6]

Washburn's description of positive feedback between tool use and bipedalism suggests that far from being an artificial "add-on," technology is an integral factor in our biology, without which it would be impossible even to *conceive* the "human." Technology is not itself a tool or artifact, but the conditions of possibility—in both body and brain—of tool manufacture and use.[7]

The peculiar relation between natural selection and conscious artifice is responsible for "the process by which, as Theodosius Dobzhansky says, biological evolution has transcended itself."[8] And it is in light of this genetic intertwinement that technology first presents a "problem," namely how the artificial is to be distinguished from the natural. There is no question that there is a *phenomenological* difference between the natural and the artificial that is intuitively obvious to everyone. Thus Leo Lionni speaks of "organicity, a basic quality common to everything in nature, and the one that usually sets an immediate and unmistakable stamp on outward appearance,"[9] which "typifies the forms of nature and which is lacking in the products of man."[10] But in what this organicity (and its opposite) consists is a puzzle.

In *Physics* II, Aristotle distinguishes the natural and artificial beings in terms of their respective "principles of motion": the former contain this principle within them, the latter are moved from without.[11] While this criterion is useful for distinguishing a stone, plant, or animal from a bed or toothbrush, it does not help us to understand artifice itself as part of human *nature*. For while an artifact is "moved" (i.e., manufactured, used) by an outside agent, namely the human being, the human *naturally*—that is, by his nature—makes and uses it.

We might therefore say with Aristotle that an artifact is a *product* that has *come to be* as a result of *craft* (*technê*), where craft is

the study[12] of how something that admits of being and not being comes to be, something whose origin is in the producer and not in the product. For a craft is not concerned with things that are or come to be by necessity; or with things that are by nature, since these have their origin in themselves.[13]

Craft, *technê*, is "in the producer": that is to say, technology is in the human being as that by which[14] beings are brought into existence that would not otherwise have done so, viz., "by nature" or "by necessity."[15] This phrasing—"neither by nature nor necessity"—suggests that technical objects find their origin in a moment of *free* creation. It is just in this freedom that the "artificiality," the "*man*-madeness" of artifacts is rooted—we shall return to this point below.

The fundamental insight that "technology" is not in the first place outside of but in us was clear to Aristotle and Plato. Yet it is obscured by our common way of speaking of "technology" as a real thing: indeed, equipment, tools, utensils are often simply called—loosely and inaccurately—"technology." Such talk is not merely a matter of linguistic sloppiness; it has its root in a phenomenological obscurity inherent to technology itself. Precisely *because* the human being is naturally and constantly involved with tools, his attention is directed outwards towards those tools in their materiality, and away from his inner, immaterial, technological nature.

But in just this "externality" of the product with respect to the producer lies another sense in which technology is a "problem." The craftsman or producer, the *technitês*, by craft, brings forth (*pro-ducere*) his product, the artifact, which now sits there before him as something *thrown-forth*: quite literally, his *pro-blêma*. But why should the product be a problem for the craftsman? Because in *crafting* the product, he was free, but as soon as the product is finished, the freedom has passed, and the product now returns to the realm of "nature and necessity." It stands there as a testament to, a reflection of the craftsman's freedom—but of his *past* freedom, manifested and expressed in his technical *activity*. When his activity ends and simultaneously issues in the product, his freedom, too, passes from him altogether.[16] The artifact is the material expression and proof of the craftsman's free activity of *technê*, but insofar as it stands over against him as "finished" and so in a sense *autonomous*, the maker is *alienated* from his product.

Of course, the craftsman does not simply make a product to let it lie there: it is made to be *used* or *consumed*. His craft is implicated in an economy of crafts, his product in an economy of products. Again, we can look to the ancients for the simplest and clearest statement of this fact. In *Nicomachean Ethics* I.1-2, Aristotle seeks to locate the highest good. To find the highest good, Aristotle notes that "every craft [*technê*] and every investigation, and likewise every action and decision, seems to aim at some good"; and where the end lies beyond the action (i.e., where the activity is not itself the aim, as in taking a walk or playing the piano), "the product is by nature better than the activity."[17] Each end is aimed at by a particular craft[18]; these ends are not all equal, but are organized hierarchically,[19] namely with respect to some ultimate end. For example, "bridle-making and every other [craft] producing equipment for horses are subordinate to horsemanship, while this and every action in warfare are in turn subordinate to generalship, and in the same way other [crafts] are subordinate to further ones."[20] Aristotle here outlines a hierarchical economy of products, and thus of the crafts that produce them. The bridle is *for* riding horses, and horsemanship is *for* victory in battle; hence a bridle is subordinate to horse-riding, which is in turn subordinate to fighting on horseback, just as the craft of bridlemaking is subordinate to horseman*ship*, which in turn is subordinate to the *art* of the cavalry battle. Now, what Aristotle here calls the relation of subordination may just as well be described as the relation of use or consumption. The rider *uses* the bridle, and thus the bridlemaker, whereas the bridlemaker does not use the rider. Instead, the bridlemaker uses the product of the tanner or the smith, whose crafts are in turn and for this reason subordinate to his.

If in Aristotle we find a hierarchical economy of crafts and artifacts, Plato imagines a "flat" economy, in his theoretical generation of the so-called city of pigs in *Republic* II.[21] There, Socrates also locates the origin of human community in an economy of crafts. Humans need crafts to counterbalance their congenital frailty. My feet are soft, so I need shoes; I have no fur, so I need clothes; I cannot find enough to eat, so I must farm; etc. But if I try to make my own shoes, clothes, shelter and food, I will end up doing none of them well, and likely fail in all. For this reason, I (and every other person) has cause to join in an economic community of specialists, each of whom supplies the demands of all the others and himself. This community is marked by a dynamic equilibrium of supply and demand; no more is produced than is consumed, and no more is consumed than is necessary to compensate for the natural weakness of each of the community's members.

The difference between Aristotle's hierarchical and Plato's flat economies is rooted in the difference in their respective ends: the Aristotelian ends lie outside the economy, whereas the end of the city of pigs is nothing more than its self-preservation. It has no "values" or "aims" over and beyond its own existence, that is, the preservation of its inhabitants' lives. So, while in both economies crafts and products form systematic networks, it is only in the Aristotelian one that the worker is alienated from his work and product: for the *end* of his work is not *his own*. By contrast, in the city of pigs, each craftsman is entirely the master of his work, which he in the first place *uses and consumes himself*, and, in the second place, *immediately exchanges* for some other necessity of *his own* survival with another who will also *immediately consume* the object of their barter. This observation leads to a further distinction. Because in the city of pigs products are used either immediately or (at most) after a single exchange based on immediate need, the supply and demand are in perfect equilibrium; the community's economy lies "flat" upon the basic plane of need. The community governed by "values" and ends over and beyond immediate need, by contrast, are in a perpetual state of disequilibrium, subject to conflict and competition among "values,"[22] and, if the community is motivated (at least in part) by the value of luxury or pleasure, to the indefiniteness of the thirst for *more*.

In nuce, the city of pigs is the city of "enough," while the city of "values" is the city of "ever more." It is in the former that the craftsman is least alienated from his *technê*, because he is his own master. Moreover, because in the city of pigs artifacts are produced solely on the basis of immediate need, they are also immediately consumed or put to use, so that this economy builds up no *store* (*Bestand*) of products. It is likely in light of this dynamic equilibrium of need and satisfaction that Socrates calls the city of pigs the "healthy" city. The city of luxury (which is also the city of pleasure, honor, and philosophy), is "feverish," for it is full of unnecessary and therefore disruptive crafts and products.

It is interesting to note, then, that while both communities are at root technical economies, the one is healthy because its *technai* stand in a natural relationship to human *nature*, both biological and spiritual. The other city is sick, not because it is "technological," but because its technologies have swung out of balance by the conflict of "values"—that is, by notions of what a city needs to be fit for something "higher" than pigs.

Technology is thus again a "problem" when its *natural economy* is disturbed by the devising of *artificial ends*, towards the achievement of which the economy is now steered. What Socrates touches on in such a fleeting way, then, in calling the city of pigs "healthy" and "true"—only to follow Glaucon's lead in delving deeply into the problems of the deluxe economy—finds its most sustained elaboration in the Chinese classic, the *Tao Te Ching*.

The Taoist is fundamentally opposed to all value-systems *as such*. Historically, his immediate opponents were of course the Confucianists with their fancy system of hierarchies, rites, and "music," all of which were meant to make life *humane*—that is, something above the beast. But in this very notion of *human value*, the Taoist finds the seed of human corruption. The main thrust of Taoism is the natural balance that exists between opposites.

> The way of Heaven—
> Isn't it like stretching a bow?[23]
> You press down the high,
> Raise the low,
> Take away the excess,
> Add to the deficient.
> The way of Heaven

Takes away from those who have too much and gives to those who havenot enough.[24]

The introduction of any artificial "end" or "value"—even (or especially) one as noble-sounding as "humanity"—necessarily leads to a disruption of the equilibrium that is the *Tao*: "The way of man/Is not so."[25] It is in this light that I understand Lao Tzu's constant admonition against "doing" or "acting," and his correlative approbation of *wu wei* or "non-action." To refrain from "action" does not imply a negative quietism, but rather, acquiescence in the *Tao* as the *given* economy of nature. To take "no action" is thus not in the first place a stillness of the body, but a stillness of mind—what the Skeptics call *ataraxia* or "unperturbedness"—by which again I do not understand a vacuum of thought, but a refraining from *committing* oneself to artificial values and ends; for these can only serve to obscure the "uncarved" clarity of the *Tao*.

To say that the human being is "by nature" technical should not be construed as meaning merely that he naturally makes or uses tools—at least so long as his "making" and "using" are not correctly understood, an understanding that involves a peculiar difficulty. The character of our technical nature is obscured precisely when it is thematized, that is, just when we turn our attention to this nature itself as an object of contemplation. It is like the phenomenon of perceiving something from the corner of one's eye, only to have it disappear in the blind spot when one directs one's gaze at it. This is because in thematizing "technology," we tend to focus on tools or equipment as things "lying there" before us, awaiting our use. But in doing this we do not see our tools as they *are*, that is, in their *essence* as *things in use*; rather, we see them in a distorted fashion, as "objects of technology." For example, when I consider my pen as an example of technology, it does not appear to me, the theorizer of technology, as a *pen*—an appearance that is reserved for the unselfconscious practical *user* of the pen. An extreme if telling analogy is this: if I were to pull out a tooth and examine it with scientific detachment, I might learn a great deal about it—its shape, weight, chemical composition, etc.—indeed, more than if I had left it in the mouth. But the whole time it lies there before me, it is not a tooth in the strict sense, that is, insofar as *being a tooth* is to *perform the tooth's functions*. So, too, a pen and any other tool or "piece of technology" *is* only *as* it is *in use*. But when I am *using* my pen, I cannot easily contemplate it, at least not from a first-person point of view, for to use an implement is generally not to be explicitly or thematically aware of it *as* "implement": rather, one is aware of it as an organic or integral element of an activity that one is engaged in. The batter will miss the ball if he abstracts himself (that is, distracts himself) from the activity of batting in order to contemplate the ball as ball, the bat as bat.

Therefore, to see tools in their whole context, we must contemplate them from a third-person perspective, and observe the tool as it is being unselfconsciously used by another, and with whom it constitutes a single dynamic system. Only in this way can the *artificiality* of thematizing the artifact be kept from blocking our access to the *naturalness* of human manufacture and use of tools. For *usually* (*zunächst und zumeist*), tool use is not directed towards a theoretically explicit end, so that tools also do not usually appear to us as "means." Of course, the pen is "for the sake of writing," the basketball or basketball net "for the sake of the game of basketball" or "for the sake of winning at the game of basketball," but again, it is the odd writer or basketball player who makes these ends explicitly clear to himself each time he picks up a pen or a ball. These ends rather form a background of significance that any tool user could, if pressed, bring to explicit articulation: but, as I have argued, such articulation would at once distort the true relation of the tool to its background, since now the background has been brought into unnatural prominence. The naturalness of tools is most manifest in our everyday unthinking and inexplicit involvement with tools and equipment, as Heidegger most notably describes it in *Being and Time*. There, he suggests that conscious manipulation of tools indicates a stall or disruption of the smooth flow of—not "my" employment of this here tool—but of the single integral system constituted by user and tool.

Of course, it is a now again a *problem* to see oneself and one's tools as dynamic systems, for it involves contemplating tool and human as distinct or distinguishable *parts* of a whole, and then together *as* that whole, simultaneously. This very unnatural and difficult attitude is, as I said, perhaps more easily adopted by observing the human-artifact "whole" from a third-person perspective, though one must eventually also feel that unity oneself through, if not theoretical, then mindful attention to one's own tool-using *activity*. Nevertheless, my point is not hard to see, given a few simple examples. Instead of saying: "She sits in the chair," I consider the single "sitting-system," which consists in what I am wont (by linguistic habit) to see as two things, subject and object, the sitter, and the sat-in chair. Again, instead of saying, "The man is driving the car," I consider the driving-system, that is, *driving*; for the man is not a driver without the car, nor can the car drive itself, that is, *be driven* without the man. Another way of putting the point is to consider the chair or car as extensions of the human being, not in the sense (merely) of a prosthesis (for this maintains an obscuring distinction between user and instrument)—but in the sense that for the time I sit in a chair or am driving the car, an entirely new *entity* comes into being: the "chair-man," or the "car-man." As odd as such terms sound, daily life would seem to consist largely in moving among different such systems, becoming successively or simultaneously a "bed-man," a "clothes-man," a "table-chair-man," a "house-man," a "car-man," a "telephone-desk-chair-computer-man"—a mundane, if no less powerful descendant of the centaur-archer.

These entities, into countless of which each of us continuously metamorphoses each day, are not, despite their artifactual elements, *artificial*: they are the *natural*, because *essential* forms of human being. Now, I would argue along similar lines as before that where a "problem" may arise in the relation between the human being and his tool is not in the simple fact that she is using a tool: she *is* the tool user, for each given tool (and when she is not using a tool, she is defined precisely by her *lack* of a tool—in a way that no other animal ever can be). Rather, the "problem" again lies in an *imbalance* that arises between the user and her instrument, an imbalance that originates just in this very distinction (sc. between user and instrument).

The natural balance that constitutes the integral unity of (to speak conventionally) user and instrument is (like) the natural economy of tools discussed earlier, but at the level of a single agent. Just as I argued that such a natural economy is upset by the introduction of artificial values or ends, so too the balance between "me" and "the tool" is destroyed when I consider the end towards which "I" am employing "the tool"—an end which in some sense must always be impertinent to the using activity itself. Consider again a basketball or piano player. The "end" of *playing* basketball or piano is the playing itself: "for the love of the game," "for the love of music"—which must mean, for the love of *playing* basketball or *playing* piano. When the player *is* a player—that is, merely playing for its own sake—the unity of the instrument (the ball, the net; the piano), on the one hand, and the player's body and mind, on the other hand, is inviolable. When the innocence of mere play is overlaid by the value of "performance" not to speak of "winning," the ball or piano *now* become "means," viz., towards "excellence" or "victory."[26] Only in this way are phenomena such as "choking" or "stage fright" explicable, for the player lost in unselfconscious practice—that is, unadulterated *praxis*—of his art cannot "lose."

Now, although in lifting my glass to my lips to drink, or switching on the light as I enter a room, I act unconsciously (sc. of any explicit goal) and therefore form a perfect if fleeting "glass-man" or "light-switch-man," effortlessly quenching my thirst, or illuminating the dark, yet just when it comes to "important" matters—that is, "ends" that I "value"—I become self-conscious, my actions labored. My opinions of good and evil directly condemn me to toil among the thorns and thistles instead of living a life of easy perfection. That is, it is through values that I become alienated from the tools that are *parts* of my human existence.

It is in this light that we might also understand the peculiar fact that such practices of enlightenment as Zen meditation or yogic contemplation involve not the abandonment of technology but on the contrary consist in elaborate techniques of control. Consider the title of Eugen Herrigel's book, *Zen in the Art of Archery*. Archery is an art, that is, a technique of using a tool, the bow and arrow. The consummate user of this tool is the "master" of its use—and this is none other than the "Zen Master." Now, the point of Herrigel's book is that so long as the "art" of archery is conceived solely as the manipulation of the bow and arrow to achieve a desired end, the archer will remain mastered by that end, and so too remain alienated from his instrument. The genuine art of archery—or any other art—just is the universal *Zen technique* of overcoming attachment to and desire for any end whatsoever. For as soon as ends and values are transcended, their distorting effects upon the economy of mind and action are dissolved, and the practitioner (the erstwhile "user") comes to be "at one" with his tool, that is, reconstitutes the integral system of "man-instrument" in its original dynamic state of play.

A unified entity results from or supervenes on the equilibrium between human and machine, an equilibrium that naturally obtains when ends are abandoned and tools consequently cease to appear as means to such ends. For it is ends—*artificial* ends—that cause us to become conscious of our instruments as means, and of ourselves as users "of" these; and it is in this self-consciousness that the alienation I alluded to earlier consists.

Now the final problem that arises is this: can one enter into an "endless" relation with just any tool or machine, or are there some technologies or techniques that are essentially alienating and alienated from the human being? Lao Tzu and Socrates seem to agree that the latter is the case. According to the Taoist, we humans have a tendency to innovate, to try to improve upon nature; this nature is the "uncarved block" of the *Tao*, whereas our values, ends, and actions promoting those values and ends are the carvings upon that block. This abstract and "mysterious" doctrine takes on concrete meaning in the city of pigs: this community is "uncarved," its inhabitants living in harmony, in accordance with nature. From the point of view of Glaucon, the civilized man of "higher" values, they live like pigs; but this is because the civilized man *by definition* stands opposed to nature. Nature is merely the block, the foundation upon which he inscribes his values. In this sense, the rest of the *Republic* may be seen as a carving, which, the more elaborate it gets, the more it obscures the original health and harmony of the *Ur*-community that once underlay it.[27] Thus all of the civilized *technai* of the city of luxury cannot help but be alienating. The city will need the crafts that provide

couches, tables, and other furniture ... and, of course, all sorts of delicacies, perfumed oils, incense, prostitutes, and pastries[;] ... painting, [and] embroidery[;]

and craftsmen, including

hunters, artists, ... [musicians], poets and their assistants, actors, choral dancers, contractors, and makers of all kinds of devices...[; and] servants, tutors, wet nurses, nannies, beauticians, barbers, chefs, cooks, and swineherds[, and] doctors[, and] a whole army.[28]

Within a given craft it would seem possible to attain a balance between user (or maker) and tool (or object-being-made); after all, the pianist, the Zen archer, and the basketball player[29] are all manifestations of the city of luxury. Nevertheless the economy of *technai* as a whole will be out of balance, because, at least on Socrates' account, there will always be at least two competing but mutually dependent socio-political factors at work, namely the consumer-industrial complex and the military-industrial complex. The latter is made necessary by the insatiability of the former. But where the epicure is motivated by pleasure, the guardian aims at honor; and as Plato, Aristotle, and Confucius all seem to agree, it is to resolve the resulting conflicts that the *philosopher* enters the scene, superimposing yet another layer of "values" upon the palimpsest that was once the uncarved block of the primitive community.

Thus the inhabitants of the luxurious city will necessarily experience *technological* alienation. For whether the whole city is organized around just one highest value or end—pleasure (Sodom) or honor (Sparta)—or suffers an inner conflict among ends, the city's principle(s) will not be congruent with the ends of all or even most of the people's. As Aristotle's description of a command economy in the *Nicomachean Ethics* I.1-2 suggests, not just the bridle but the bridlemaker will be part of the vast machine that is the state, serving the statesman's end. Similarly, the producer in the Platonic *kallipolis* will take his orders from a superior, regulating his activity in accordance with the philosopher's vision of the Good—a vision *ex hypothesi* unavailable to the lowly worker. By the same token, the philosopher or soldier who is the citizen of a pleasure-oriented city will feel apart and a stranger in his own state. So no matter how "integrated" an individual may be in her own craft, she will always exist as a mechanism in the engine of state, serving ends she may or may not even be aware of or approve.

Such a situation is impossible in the city of pigs precisely because there is no hierarchy of "values." Its inhabitants presumably feel pleasure—indeed their life is purely pleasurable (*Rep.* 372a-d)—and perhaps even, as Rousseau describes the simple village life that in many ways echoes that of the city of pigs,[30] have a primitive sense of "merit," "consideration,"[31] and even "justice."[32] The critical difference is that these *natural feelings* are not for the savage pounded and processed into the "value-artifact," an abstract object distinct from him, and thus an idol of his own making before which he now bows down as something "objective." His life is pleasant, but not governed by pleasure; he recognizes others and is recognized, but has no ideology of honor; and he does his own job without meddling with the others', and so, in the sense of the term peculiar to the *Republic*, enjoys "justice" without making it a notion of political theory. In all these ways, he fails to make pleasure, honor, and wisdom into objects of technique; *a fortiori*, he fails feverishly to strive after the images of things he already, in accordance with nature, possesses. This failure to strive—the "health" of the city of pigs—is none other than Taoist non-action (*wu wei*).

One of the most important theorists of technology is Heidegger. In his *Die Frage nach der Technik* ("The Question Concerning Technology"), he identifies the essence of modern technology as the so-called *Gestell* or "framework."[33] *Gestell* is, as he puts it, a mode of "unconcealment," that is, a way in which the world appears to the human. The framework, the essence of modern technology, unconceals or reveals the world as "standing-reserve" (*Bestand*), that is, resources or stock.[34] Although Heidegger considers this statement to be metaphysical, insofar as it determines the "essence" of technology, yet it is more illuminating as an economic thesis. He in effect identifies the peculiar nature of modern technology as the imperative of treating all beings in the world as resources for potential exploitation. *Gestell*'s insidious nature lies in the obviousness of this view of nature: the very banality of such terms as "natural resources" or "human resources" makes them the common coin of capitalists and communists, of industrialists and environmentalists. No one debates the appropriateness of the "resource-exploitation" interpretation of nature; they simply debate how nature is best to be exploited: for private profit or for the common good. Even the environmentalist who argues for the "preservation" of nature *ipso facto* sees it as a resource.

The danger of modern technology for Heidegger thus lies not in the threat of nuclear annihilation, environmental catastrophe, the engineering of human beings, or totalitarian surveillance and control of the population. Rather, he sees all these dangers as symptoms or effects of modernity's "world-picture," in which nature and people are seen finally as objects of technical control. Towards the end of his essay, Heidegger darkly suggests that modern technology conceals within itself a "saving power," and that this saving power is *art*. If I understand his mystifying prose[35] correctly, technology "harbors in itself ... the possible upsurgence of the saving power,"[36] because it *shares* in the original Greek concept, *technê*, a common root with art. Hence, by *questioning* the nature of technology, one might return to that root and find there an alternative way of revealing the world, viz., one that strives not to "challenge forth" the resources of nature, but to "let beings be." Art, like craft, is a mode of *poiêsis*, of making—and as I have argued from the outset, the human being is *essentially Homo faber*, or "poetic man."[37] We cannot help but encounter the world in a way that is *somehow* "technical."

I concur with Heidegger that modern technology must be questioned. But his view that salvation (from what?) lies in *art* would seem itself to harbor the dangerous seeds of new *ideologies*, and thus new and unheard-of modes of alienation and exploitation, especially if combined with the modern techniques they originally meant to challenge. Even the very philosophical and downright Socratic view that "questioning is the piety of thought"[38] could be constructed into a new and fruitless ideology of mastering the technology through ... techniques of thinking. Heidegger seems aware of these possibilities of misconstruction; for example, he cautions against interpreting "art" as "sheer aesthetic-mindedness."[39] But it strikes me that he has not sufficiently considered the intrinsic connection between technology and economy, or that technology and economy are in a deep sense synonymous.

Indeed, I would argue that once this connection is made, the "saving power" is seen not to lie in ("fine") art but in the "economic" insight of Lao Tzu and Socrates, viz. that all notions of "fineness" immediately and necessarily lead to a disruption of the natural balance of "Stone Age economies," as Marshall Sahlins calls them.[40] In closing, let us reconsider Verse LXXX of the *Tao Te Ching* from this perspective. He writes: "Reduce the size and population of the state."[41] Ha Poong Kim may well be right that the Lao Tzu's immediate concern in this proposal is counteracting the struggles among large ambitious polities during the Warring States Period (403-221 BCE).[42] But it is no less true that reducing the state is a first step towards restoring the balance between human beings and nature. In the global ubiquity of states and their populations, "man exalts himself to the posture of lord of the earth…[;] it seems as though man everywhere and always encounters only himself";[43] he no longer can recognize or even imagine what "living in accordance with nature" might mean, because "nature" itself is conceived solely as the object, the recipient of our *doings*. Lao Tzu immediately turns to technology:

> Let the thousand contrivances go idle.
> Let the people take death seriously and not move to distant places.
> Though they may have boats and carriages,
> They will not ride in them.
> Though they may have armor and weapons,
> They will not display them.
> Let the people return to the practice of knotting cords.[44]

What does this mean? In light of my preceding discussion, it is clear that Lao Tzu is not advocating a form of Luddism—for Luddism again entails a commitment to a set of values, an anti-technological *ideology*: smashing looms—or beating swords into plowshares in the name of "peace"—is a *doing*, contrary to the Taoist notion of *wu wei*. "Letting the thousand contrivances go idle" means just that: turning our backs on them. But he urges this *not* on the basis of some new pacifist or ecological table of values, but because for the Taoist all tables of values are to be—not smashed—but ignored: for they are all, as tables, artificial. Instead, *death* is to be taken seriously by the people. For once artificial, higher values—values one likes to say are worth *dying for*—are tossed aside, the human being will return to the sole and overriding concern with which he has been endowed by *nature*, namely *living*. When a person is confronted with imminent death, through illness or accident, for example, the world of "boats and carriages," of space shuttles and iPhones appears suddenly irrelevant, an absurd joke. The time spent caring for these artifacts as if they mattered seems utterly wasted.[45] If one were to "take death seriously," confront one's radical finitude, as Heidegger says, one would naturally leave aside sleek boats and carriages as trivialities, let glorious "armor and weapons" rust away as so much trash.

Moreover, Lao Tzu is not "anti-technological" in the sense of advocating a back-to-nature ideology. Not only will the people use the "knotted rope,"

> They will relish what they eat,
> Find their clothes beautiful,
> Be content in their homes,
> Delight in their customs.[46]

Human beings are makers and users of tools: they will farm or hunt, make clothes, and build homes—just as Socrates, too, describes in *Republic* II. And they will do *only* this so long as they refrain from forging values, concentrating instead on the constant presence of suffering and death. As Socrates adds, "They'll enjoy sex with one another but bear no more children than their resources allow, lest they fall into either *poverty or war*."[47]

> Lao Tzu concludes with a final economic observation:
> States may be within sight of one another,
> So that one may hear cocks and dogs from a neighboring state;
> Yet people will grow old and die
> Without trafficking with another state.[48]

The Taoist "state" suddenly appears as no more than a small village, perhaps a few hundred yards from the next. With the abandonment of all ends except that of survival,[49] its people have lost all motivation to trade or traffic with their neighbors, who, for their part also lack such motivation.[50] These communities are entirely self-sufficient, "altogether complete in [themselves]," to paraphrase Rousseau again.[51]

Lao Tzu is not prescribing a course of action that a Taoist ruler ought to take, so much as he is simply describing the natural consequences that would follow upon such a ruler *refraining* from the imposition of values on the populace, thus letting them settle of their own accord in the *Tao*, the "Way of Heaven."[52] And while the advent of such a sage may indeed be a pipedream, the anthropologist, Marshall Sahlins, shows that a state answering to Lao Tzu or Socrates' description exists (or until recently existed), namely in Paleolithic societies.[53] Such peoples as the !Kung or native Australians live, as Sahlins describes them, lives of "affluence without abundance"[54] in the sense that their (very simple) needs are in equilibrium with their ability to satisfy them. At the same time, they are *masters* of technology and techniques, making and deploying tools *precisely* as and when the situation demands. When the moment passes, the tool is treated in just the "Taoist" way I described above:

They don't know how to take care of their belongings. No one dreams of putting them in order, folding them, drying or cleaning them, hanging them up, or putting them in a neat pile.... The European observer has the impression that these [Yahgan] Indians place no value whatever on their utensils and that they have completely forgotten the effort it took to make them.... The Indian does not even exercise care when he could conveniently do so. A European is likely to shake his head at the boundless indifference of these people who drag brand-new objects, precious clothing, fresh provisions, and valuable items through thick mud, or abandon them to their swift destruction by children and dogs.... Expensive things that are given them are treasured for a few hours, out of curiosity; after that they thoughtlessly let everything deteriorate in the mud and wet. The less they own, the more comfortably they can travel, and what is ruined they occasionally replace. Hence, they are completely indifferent to any material possessions.[55]

The European "shakes his head" and thinks, like Glaucon, that these people live like pigs, "letting everything deteriorate in the mud and wet." What really puzzles him in the Yahgan Indians' "boundless indifference" is their lack of any *value* placed on their utensils. But this very indifference indicates the complete "integration" of the Paleolithic human with the tools needed for survival. Whether a utensil is "brand-new," "precious," "valuable" or "expensive" is entirely impertinent to its utility for the task of self-preservation. If it is useful, it is *immediately* at and in hand, as if it were a part of its user. And indeed, as I argued earlier, it *is*, insofar as the user and his utensil constitute a single systematic entity. Such an integration of human and tool is made possible precisely by the user's innocence of whatever scheme of ends in which the *tool itself* is seen as "valuable." The Paleolithic aim of self-preservation demands movement, which in turn gives rise to the natural "policy" of having few things: only "what they can comfortably carry themselves."[56]

The Paleolithic human being is therefore very wrongly seen as technologically undeveloped. On the contrary, he manifests the *perfect* integration of human and tool in a vacuum of artificial values. It is we who are technologically overdeveloped. The chief problem of technology is therefore not how to master uncontrolled technical development, or how to reverse that development. The problem is what, given our present state of technology, an "equilibrium" between man and machine even *means*.

Addressing this question is of great importance if we are ever to discover what Heidegger calls a "free relation to technology."[57] We have reached, through a circuitous route, the conclusion that this is as much an economic as a "metaphysical" issue. Glaucon's scorn for the healthy city's poverty is the proximate impulse of the indefinite proliferation of technologies and techniques, and thus of the human being's alienation from her own technical essence. Sahlins expresses a similar point:

We are inclined to think of hunters and gatherers as *poor* because they don't have anything; perhaps better to think of them for that reason as *free*. "Their extremely limited material possessions relieve them of all cares with regard to daily necessities and permit them to enjoy life." (Gusinde, 1961, p. 1)[58]

I would only add: the hunter and gatherer is free of the burden of civilization's "material culture"—that is, the sum total of its *technai* and artifacts—because he is free of the thought that these are good. *His* artifacts and utensils do not surround him as the symbols and mementoes of alienation. He drops them when they get too heavy, and so remains free.

References

Aristotle. 1985. *Nicomachean Ethics*. Irwin, T., tr. Indianapolis: Hackett.

-----. *Physics*.

Catlin, G. *Wi-Jun-Jon (An Assinneboin Chief)*. 1844. Museum of Nebraska Art, University of Nebraska, Kearney, Neb. 25 Aug. 2009, http://monet.unk.edu/mona/artexplr/catlin/aecg3.html.

Cooper, J.M., Hutchinson, D.S., ed. 1997. *Plato: Complete Works*. Indianapolis: Hackett.

Ellul, J. 1964. *The Technological Society*. New York: Vintage.

Gusinde, M. 1961. *The Yamana*. 5 vols. New Haven: Human Relations Area Files.

Heidegger, M. 1993. *Sein und Zeit*. Tübingen: Niemeyer.

-----. 1977a. *Basic Writings*. Krell, D.F. ed. New York: Harper & Row.

-----. 1977b. "The Question Concerning Technology." Lovitt, W., tr. In Heidegger, 1977a: 287-317.

Herrigel, E. **XXXX**. *Zen in the Art of Archery*. Vintage Books, 1999.

Kim, A. 2009. "Animal Farm: The City of Pigs as an Ideal." Unpublished manuscript.

-----. 2005. "Frameworks and Foundations." In *Angelaki* (Special Issue: "Continental Philosophy and the Sciences: The German Tradition") 10:1 (April): 201-18.

Kim, H.P. 2002. *Reading Lao Tzu: A Companion to the* Tao Te Ching *with a New Translation*. Xlibris.

Lao Tzu.[59] 1963. *Tao Te Ching*. Lau, D.C., tr. Harmondsworth: Penguin.

-----. 2002. *Reading Lao Tzu: A Companion to the* Tao Te Ching *with a New Translation*. Kim, H.P., tr. Xlibris.

Lionni, L. 1977. *Parallel Botany*. New York: Alfred A. Knopf.

Montagu, M.F.A., ed. 1962. *Culture and the Evolution of Man*. New York: Oxford.

Plato. 1997. *Republic*. Grube, G.M.A., Reeve, C.D.C., tr. In Cooper, 1997.

Rousseau, J.-J. 1984. *A Discourse on Inequality*. Cranston, M., tr. Harmondsworth: Penguin.

Sahlins, M. 1972. *Stone Age Economics*. New York: Aldine de Gruyter.

Washburn, S.L. 1962. "Tools and Human Evolution." In Montagu, 1962: 13-19. Originally published in *Scientific American* 203 (1960): 63-75.

Notes

1. Rousseau, 1984: 82.
2. Cf. Rousseau, 1984: 68, 78.
3. Catlin, 1844.
4. Washburn, 1962: 13.
5. Washburn, 1962: 13.

6. Washburn, 1962: 16.
7. Cf. Heidegger, 1977: 287: "Technology is not equivalent to the essence of technology"; "the essence of technology is by no means anything technological."
8. Washburn, 1962: 14.
9. Lionni, 1977: 35.
10. Lionni, 1977: 37.
11. Aristotle, *Physics* II.1, 192b.
12. [GREEK]
13. Aristotle, *Nicomachean Ethics* VI, 1140a.
14. I purposely avoid terms such as "capacity" or "power" here: *technê* is an essential mode of human *be-ing*: to call it a capacity would again be to externalize it as something "detachable" from the human.
15. Note: a) thus craft is an expression of human *freedom*; b) this definition is broad enough to include as "products" things that are "useless"—the *pure* expression of this freedom.
16. → economy (NE 1.1)
17. *NE* I, 1094a.
18. Departing here and below from Irwin's "sciences" (1094a10); the Greek simply says "as many as there are of such," viz. the aforementioned medicine, boatbuilding, generalship and household management; the example that follows, bridlemaking, shows that "craft" is a more appropriate translation here.
19. Ibid.
20. Ibid.
21. I treat the political economy of the City of Pigs in detail in "Animal Farm: The City of Pigs as Ideal" (2009).
22. The three basic contested values being, according to Aristotle, pleasure, honor, and contemplation.
23. Cp. Heraclitus, Fr. 51.
24. *Tao* 77:1-8 (tr. Kim).
25. *Tao* 77:9-10 (tr. Kim).
26. Cf. Ellul, 1964: 382, ff.
27. Indeed, as I argue in my 2009 essay, the institution of the guardian-class can be seen as an attempt to reestablish the harmony of the city of pigs at a higher social level.
28. *Rep.* 373a-e.
29. Cf. Ellul, 1964: 382, ff.
30. A key difference is that Rousseau believes the villagers of this "happiest epoch" (Rousseau, 1984: 115) restricted themselves to "work that one person could accomplish alone and to arts that did not require the collaboration of several hands," being drawn into community by the pleasure of each other's company (Rousseau, 1984: 113-4). Plato by contrast supposes that community arose precisely from "one man need[ing] the help of another"—a situation that Rousseau thinks led to the introduction of property and exploitation (Rousseau, 1984: 116).
31. Rousseau, 1984: 114.
32. Rousseau, 1984: 115.
33. Heidegger, 1977b: 301. I do not follow the English translator in using the barbarism, "the enframing."
34. Heidegger, 1977b: 302.
35. E.g., Heidegger, 1977b: 312-7.
36. Heidegger, 1977b: 314.
37. Our ancestors after all included *Homo habilis* (handy or expert man) and *Homo erectus*, whose very posture was, as Washburn argues, determined by his carrying and manipulating tools.
38. Heidegger, 1977b: 317.
39. Heidegger, 1977b: 317.

40. See Sahlins, 1972.
41. *Tao* 80:1 (tr. Lau); Kim: "Let the state be small and its people few"—a similar sense if an entirely different tone; see below.
42. Kim, 2002: 166-7. Cf. *Rep.* 373e, ff.
43. Heidegger, 1977b: 308.
44. *Tao* 80:2-8 (tr. Kim).
45. Heidegger describes this phenomenon somewhere in *Being and Time*.
46. *Tao* 80:9-12 (tr. Kim).
47. *Rep.* 372bc, emphasis added.
48. *Tao* 80:13-16 (tr. Kim).
49. "Self preservation [is] the savage's only concern..." (Rousseau, 1984: 86).
50. In this respect, Lao Tzu and Socrates part ways, as the latter believes trade to be necessary even for the primitive polity; see *Rep.* 370e, ff.
51. Rousseau, 1984: 82. Rousseau argues along similar lines for the natural *un*sociability of "natural man"; cf. e.g., Rousseau, 1984: 84-6.
52. *Tao* 77:1 (tr. Kim, Lau); cp. Heidegger's notion of *Gelassenheit*.
53. His book, *Stone Age Economics*, is a sustained elaboration of the view that such societies constituted the "original Affluent Society" (as he appropriates John Kenneth Galbraith's famous phrase). A detailed discussion of Sahlins' points will have to await another time.
54. Sahlins, 1972: 11.
55. Gusinde, 1961: 86-7, quoted in Sahlins, 1972: 12-13.
56. Sahlins, 1972: 11.
57. Heidegger, 1977b: 287.
58. Sahlins, 1972: 14.
59. Whether I cite from D.C. Lau's or H.P. Kim's translation, I will abbreviate the title as *Tao*, separately indicating the translator.

Human Nature Unbound: Why Becoming Cyborgs and Taking Drugs Could Make Us More Human[1]

William Cornwell

Overture in a Major Key

Humans have ambivalent attitudes about technology. Tools free us from drudgery and difficult physical labor, subsistence living, disease, inability to travel or communicate widely, and vulnerability to the capriciousness of nature. Yet, we worry that modern technology alienates us from our humanity, makes us mentally and physically lazy, causes us to see nature as something to be tamed and controlled rather than appreciated and preserved, and leads to shallow and unrewarding lives lacking genuine connections to each other and to the natural world.

Contemplating a future with more, and more advanced, technology magnifies a person's appreciation of or anxiety about humanity's current relationship with our equipment. Although popular culture has had futurist cheerleaders like Walt Disney and his "Tomorrowland"—a utopia in which the solutions to all social ills lies in the rational development and application of technology—the last few decades' literary and cinematic visions of the future often have been bleak. The 2008 blockbuster animated movie *WALL-E* is only one example, with its twenty-second-century obese people who are so out of shape that they barely can walk. These helpless blobs, having evacuated Earth because of its overflowing trash, are entranced by mindless electronic entertainment and marketing as they recline in easy chairs that zip them from place to place on a spaceship.[2] Techno-nightmarish stories about the future gain their cultural resonance from the fear that as technology does more for us, we inevitably do less and consequently become infantilized and dependent. As machines and computers take over the thinking and labor for us, we are doomed to a life of softness and, perhaps, decadence, like a fading aristocracy living off the labor of others.

I do not think that optimists or pessimists about technology are entirely correct, but I will argue that virtue ethics shows that the appropriate use of tools and technology will help us exploit more of our higher-order potentialities and live a better life.[3] My argument is that we are by nature problem solvers and builders and users of tools, and we should not estrange ourselves from these, some of our highest, capacities. Furthermore, as tools minimize our involvement in more mundane aspects of physical and mental life, more of us can live a life once limited to the élite and have time and energy for socializing, educating ourselves, making or appreciating art, experiencing nature, solving problems, helping others, and enjoying additional aspects of a good life. Hence, a future that includes seemingly freakish ideas, like humans with computer chips that are implanted in brains awash in drugs and nanotechnology designed to enhance mental powers, does not have to alienate us from human nature; rather, if we make the right choices, these developments will be an expression of that nature and a foundation for developing other aspects of our humanity. Although it may sound paradoxical, even the strangest of the artificial and unnatural devices that we call "technology" can make us more human.

Virtue Ethics, Human Nature, and Natural Selection

Ethical theorizing, which is meant to tell us how to live, requires understanding the needs, aspirations, and capacities of humans—in short, understanding human nature. Aristotle, the seminal virtue ethicist, identifies two distinctive aspects of human nature: *intellectual* capacities to use language and reason to discover abstract, universal truths and *moral* capacities to control and shape our animal nature so that we can pursue long-term goals, sustain friendships and other important social relationships, and promote a just society. The proper or improper cultivation of these capacities leads to virtues (excellences) and vices (defects) in someone's character. Aristotle's analysis leaves out an important aspect of our humanity: the ability to consciously refine our bodily skills, and thus Aristotle omits the corresponding class of *sensorimotor* virtues that are manifested in crafts, arts, sports, and so forth. To develop and exercise all of the human virtues is to live the best human life, a life of happiness and rational activity.

Although the existence of human nature was self-evident to Aristotle and many later thinkers, "human nature" has been held in suspicion by some contemporary philosophers because "essentialist" theories have been used to legitimate traditional social arrangements that disempower women, minorities, and other oppressed groups, and Aristotle's disparaging remarks about slaves and women feed the suspicion that rhetoric about human nature is used to ratify imperfect, traditional social arrangements. Although one must be cautious developing theories of human nature to analyze virtues and vices, if virtue ethics is reduced to a modest and cosmopolitan core, the fundamental desideratum is a theory not of universal, highly specific actual or desirable patterns of behavior but of universal capacities and needs. On this unpretentious conception of virtue ethics, there can be varied forms of life that enable people to live well, but these forms of life are possible and desirable only because humans share underlying human and animal natures characterized by innate sensibilities, capabilities, and limitations. As Aristotle recognizes, every animal species has a characteristic and innate nature that partly explains its possibilities and limitations and that sets boundaries on what sort of good life it can lead. Humans are no exception.

It is one thing to recognize, as the ancient Greeks did, *that* we have an innate humanity, but it is something else to know *why* we have it. The answers were couched in myth and fable until Charles Darwin solved the riddle: Human nature arose through natural selection. The gene pool for *Homo sapiens* was shaped by the evolutionary imperative that species adapt to their environment or die off. Our ancestors were not exempt from this iron law of evolutionary biology, so our species' evolutionary history has stamped our bodies and minds. It is not a fluke that humans are good problem solvers or that we have an innate "universal grammar" that is a foundation for language use—these abilities have obvious evolutionary benefits.

So, if we synthesize Aristotle's ethical insights and evolutionary theory, we can say that virtue ethics is founded upon a conception of human nature constituted by a gene pool shaped by natural selection. Intellectual, moral, and sensorimotor virtues are the development of innate and adaptive human capacities of the central nervous system. To say that the central nervous system is evolutionarily adapted to do certain things (e.g., solve problems and use language) is to say that these things are *proper functions* of the central nervous system (or, to simplify the terminology here, of the human *mind*).[4] In the language of the ancient Greeks, a mind's proper functions are its *telos*, although one also might say that what the proper functions effect are the *telos*. For instance, from one perspective, thinking through a problem properly is a *telos* of the mind, but from another perspective, the *telos* is not the activity of thinking but the activity's intended product, namely, a correct solution.

Let us now integrate these concepts with the work of Aristotle and other virtue ethicists: Humans have innate mental capacities that can blossom into excellences or virtues if given the right support and cultivation. To reformulate that point in the language of proper functions, virtues, or the proper development and exercise of these mental capacities, arc proper functions of the mind or central nervous system. To live a good life is to take the raw potentialities provided to us by natural selection and refine them to their highest degree.

A distinctively human form of life cannot be understood without reference to widespread use of improvised tools. A tool is *improvised* to the extent that its creation or use is not instinctual (or, more accurately, is not instinctually manifested when given the right stimuli). Once something has been improvised, it can be copied and become part of the culture, but if its creation and use continue to depend upon culture and not be pre-wired at birth, then it remains an improvised tool, a cultural artifact. A bird's nest and a beaver's dam are, I assume, mostly non-improvised tools whose creation and use are not so much expressions of general problem-solving and learning abilities as of specific adaptive behavioral patterns. Human language is to some extent non-improvised (the normal brain has an innate "universal grammar") but to some extent improvised (the particular grammatical features of a language are not known innately, nor is the semantics of the language). What is exceptional about humans is our ability to develop and use improvised tools, "courtesy of an empowering web of culture, education, technology, and artifacts" (Clark 2003, 10).

Because our tools are improvised and are passed along culturally, technology can transform how we live and what we can do, long before it would cause significant changes to the gene pool. In other words, culture can evolve rapidly while human nature remains mostly unchanged. Thus, even if people of European stock today are not significantly different genetically from their medieval ancestors, most are substantially different people culturally, technologically, morally, economically, and spiritually. Yet, although a culture of tool-making and tool usage is capable of evolving far faster than human nature, that culture depends upon the capacities found in that relatively stable nature. David Rindos (1985, 71-72) describes well the complex interplay between (a) the general capacities made possible by our genetic nature and (b) the artificial and quickly evolving culture rooted in those capacities:

> The *capacity* for culture was the result of the action of an evolutionary process upon genes. The expression of this cultural capacity in humans is in no sense qualitatively different from the expression of any other genetically determined capacity. Numerous genes interact in any individual at any time and produce a general framework that is eventually expressed in the phenotype. Yet the *specific forms* of the phenotype are also an expression of the interaction of the genotype and the environment. Part of this environment comprises members of one's own species. Given a (genetically determined) capacity for culture, the individual *by definition* has a propensity to relate to members of its own species in a highly specific manner: teaching and learning, in all senses of the terms, become significant because of an innate capacity to perform these behaviors. Therefore, the actualization of the cultural capacity (as specific culturally transmitted traits) is little more than a highly specialized, albeit uncommon, form of epigenetic development.

Selinger and Engström (2007, 557) sum up (but do not necessarily endorse) the basic idea: "From this perspective, to deny human nature is, ironically, to deny the very anthropological possibility of cognition and language developing into the diverse forms we encounter. This diversity is dependent upon a core biological matrix that can be expressed in multiple but not unlimited ways."

The final distinction that I wish to make is between therapeutic and non-therapeutic improvised tools. A *therapeutic* improvised tool's proper function is to help restore the proper function of some natural bodily system. For instance, dentures and eyeglasses are therapeutic, improvised tools. *Non-therapeutic* improvised tools are those whose proper functions are supposed to extend one's powers beyond their typical, natural limit or to give one entirely new powers. For instance, most of us can walk but get around faster with a car; most of us can see but see farther with binoculars; most of us can lift things but lift more weight with a forklift.[5] Some of these non-therapeutic improvised tools augment powers of thought, in a broad sense of 'thought' to include perceiving, cognizing, attending to or focusing upon, remembering, controlling the body's voluntary movements, exercising willpower, and so forth. I now can define *"cognitive augmentation"* as *an increase in mental powers that results from using one or more improvised, non-therapeutic tools.*

In the next two sections I will examine various forms of cognitive augmentation that are on the horizon so that we can appreciate how odd and even freakish many of these developments may seem—computer chips plugged into the brains of cyborgs to improve their memory or computational power, or drugs administered to people to make them smarter or more ethical—and I then will argue that, appearances to the contrary, these forms of augmentation, if used properly, are not ruptures from human nature but instead are the fruition of that nature. In a surprising sense, the less human and more mechanical we may appear to become, the more human we actually may be.[6]

Intelligence Augmentation

Cognitive augmentation comes in many forms. First of all, there is *"intelligence augmentation"* to make us smarter (Cascio 2009, 96). Drugs already are used to augment intelligence. Caffeine and nicotine make people mentally sharper, and Adderall and Ritalin, whose F.D.A.-approved use is for the treatment of attention-deficit hyperactivity disorder (A.D.H.D.), are being used by some people, especially college students without A.D.H.D., to increase mental focus and alertness (Talbot 2009; Cascio 2009, 98).[7] In addition, the drug modafinil, which was developed to treat narcolepsy, not only can enable someone to stay alert for over thirty-two hours but a "University of Cambridge study, published in 2003, concluded that modafinil confers a measurable cognitive-enhancement effect across a variety of mental tasks, including pattern recognition and spatial planning, and sharpens focus and alertness" (Cascio 2009, 98). Furthermore, modafinil "works without the jitter, buzz, euphoria, crash, addictive characteristics or potential for paranoid delusion of stimulants such as amphetamines, cocaine or even caffeine, researchers say" (Garreau 2004, 56). The use of modafinil has spread from the military to entrepreneurs looking for a competitive edge (Cascio 2009, 98). Neurologist Anjan Chatterjee of the University of Pennsylvania believes that as our understanding of the brain's chemistry improves and more drugs are developed, pharmacological enhancements of cognition, which he calls "cosmetic neurology," will become as common as plastic surgery as people fear being at a competitive disadvantage to other people with enhanced intelligence (Talbot 2009).

Many forms of intelligence augmentation do not involve trying to change the brain's chemistry but instead providing more cognitive resources for the brain to exploit. For instance, language, which, as I explained in the previous section, is only partially improvised, and other representational structures such as drawings facilitate cognitive access to events far removed in time and space. In a sense, drawings and language allow one to perceive an event which one did not witness. Language also enables certain types of flexible social arrangements so that people can communicate their actions and intentions and make suitable adjustments as they jointly pursue various changing goals (Sterelny 2004, 252). The creation of language also let people share the task of memory, since what one person remembered could be communicated to another person. And with the invention of written records, the printing press, and the Internet, humans have been able to store information in enduring, public forms, thus supplementing individual memory and permitting records to be shared and archived. Some complex forms of thinking and learning became possible only with the invention of written records, and the same applies to the development of robust written mathematical systems. For instance, writing an essay or book is not a matter of first coming up with everything in the head and then writing it down. Rather, the writing process helps complex thoughts be developed and articulated precisely because one does not have to hold all of the ideas in memory. Similarly, few people can do long and complex multiplication problems in their heads, but the use of paper and pencil lets someone break the complex task into manageable subroutines whose results can be recorded and used for further operations. Some people have argued that in this sort of case, the paper and pencil are literally part of the person's mind, with the paper being an external form of memory (Clark and Chalmers 2000; Clark 2005; Clark 2006; Menary 2006). Others have argued that, at least for now, the mind supervenes only upon the central nervous system, which can make use of tools that complement or function somewhat like the mind's own cognitive equipment (Adams and Aizawa 2001; Adams and Aizawa 2008, chapter 7 ("Extended Cognitive Systems and Extended Cognitive Processing" Rupert 2004; Weiskopf 2008). In any case, the person's intellectual abilities are fundamentally transformed by intelligently using improvised tools for solving mathematical problems, writing essays, or doing other complex cognitive tasks. Somebody making good use of these non-therapeutic, improvised tools has, by any behavioral criterion, become smarter.

Computers provide remarkable intellectual augmentation. People can try to find patterns in data from economics, physics, commerce, and so forth, but computers do so much more efficiently. Computers can run simulations of events that are billions of years in the future and involve an astounding number of variables—these simulations would be impossible for humans to do in any reasonable period of time without computers.

Some forms of intelligence augmentation increase our ability to perceive and communicate. Telephones, the Internet, and so forth increase access to information and to other people, who are the greatest intelligence enhancers we have at present. Non-therapeutic devices such as telescopes, microscopes, radars, and so forth that increase our perceptual powers also augment what we know.[8]

The examples of intelligence augmentation I have given involve external devices in the broadest sense of the term to include pharmacological compounds, but a potentially giant breakthrough in cognitive augmentation is one that is just underway and about which most people are ignorant, namely, creating a mind/machine interface. We are accustomed to the mind controlling machines indirectly through the body's motor behavior: typing on a keyboard, touching a screen, flipping a switch, turning a dial, and so forth. Similarly, we are used to the mind getting information from computers and information networks indirectly through the senses. Nonetheless, *the mind/machine barrier is cracking*. Consider therapeutic, artificial retinas that let formerly blind people with damaged retinas see, although only at the crude level of 60 monochrome pixels (California Institute of Technology 2009; see also Cybersenses 2009 for information about a more primitive artificial retina that generates fewer pixels). An artificial retina does not communicate with its user through the senses; rather, it is a functioning part of the senses. In addition, even though this system is therapeutic and provides much coarser visual information than an innate, properly functioning visual system, medical technology has advanced at an astonishing speed, and it is likely that artificial retinas eventually will outperform the unaided eye. It is not far-fetched to suggest that one day we might routinely replace or supplement our eyes with artificial systems with magnification/zooming, night vision, and other standard features of many electronic devices. Given that therapeutic, artificial cochlear implant surgeries already have been successfully performed (Clark 2003, 16-17; Cybersenses 2009), it would be unsurprising if artificial ears with enhanced powers of hearing were developed. These systems would not be therapeutic; rather, they would augment our perception and hence our intelligence. We all might end up like the character Steve Austin, with his artificial implants for greater strength and perceptual acuity, in Martin Caidin's novel *Cyborg*, which was the basis for the television show *The Six Million Dollars Man*.[9]

How long will it be until implants are developed not only to correct or augment perceptual processes but also to increase our reasoning or memory capabilities? Imagine having a chip implanted in your brain to do the sorts of mathematical calculations that humans do slowly and unreliably, if at all, in the head, or having an implanted memory chip so that you would have almost limitless capacities to remember experiences, instructions, reading materials, and so on. These chips also could store information directly uploaded from computers or the Internet. Describing someone then as a "walking encyclopedia" might no longer be figurative. Perhaps, though, what will be implanted are not processors but wireless communication devices to connect us to "the cloud" of computing power that surrounds us even now. Only time will tell which approach makes the most sense or how these approaches might be combined to complement each other or to create crucial redundancies to protect against equipment failure.[10]

Another form of augmented intelligence on the horizon is "*augmented reality*," which involves integrating information and functionality into a person's perceptual experiences and thereby creating a "deliberate blurring of the boundaries between physical and informational space" (Clark 2003, 53). One example of augmented reality would be for a person to put on a pair of glasses that projects information onto the person's subjective visual field. Construction workers could wear glasses that project lines along the ground in their visual fields to mark the location of underground electric, water, and sewer lines, and "Already, researchers at the Technical University of Munich are looking at ways to display x-ray and ultrasound readings directly on a patient's body. A research project at BMW is exploring how an augmented-reality view under the hood might help auto mechanics with diagnostic and repair work."[11] Augmented reality could be done more easily and seamlessly if integrated with other forms of augmented intelligence. For instance, if someone already had artificial retinas and implanted computer chips with global positioning satellite technology and other wireless capabilities, then it would be that much simpler to have projected images that tagged buildings, streets, and people, and to have arrows projected to show directions to desired locations. Augmented reality also could involve embedding information into our perceptual experiences by using sounds or other types of sensations.[12]

Although this list of types of augmenting intelligence is incomplete, it conveys how much our society may be transformed by developments already underway in computer engineering and neuroscience. Some people might be deeply disturbed by the seeming dehumanization of people with computer chips in their heads and drugs in their veins to make them smarter. As Steve Mizrach writes this about cyborg technology: "Even those not spiritually inclined who still nevertheless possess the feeling that there is something within humanity which is not found in animals or machines and which makes us uniquely human, worry that the essence of our humanity will be lost to this technology."[13,][14] Yet, *all of the things that I have talked about build on human nature without changing or denying it. We are by nature problem solvers, and the creation and use of tools is an extension of that fundamental capacity.* The history of human culture, from its hunter-gatherer roots to its modern, globalized manifestations, exemplifies increasing human dependence upon tools of our own making. The "superhuman" future people that I described are every bit as human as we are, and the tremendous cultural changes envisioned should not blind us to our shared humanity with people of the future, just as we should not distance ourselves ontologically from people of the ancient world, even though our technologically enhanced lives today would look foreign and magical to them. So, even though, from a contemporary perspective, the lives of these intellectually super-enhanced humans might seem unnatural and monstrous to us, I do not see a good philosophical basis for holding that belief, unless one also is going to claim that people today should not be using computers, automobiles, and countless other devices that augment our agency, and if one were to make that latter claim, one would be denying a vital and distinctive aspect of human nature, namely, the ability to create and use tools of greater and greater sophistication. From the standpoint of virtue ethics, people's ability to leverage their intelligence in one domain, that is, the making and use of tools, in order to be smarter in a different domain is praiseworthy and accords with Aristotle's dictum in the *Metaphysics* 1.1 that all people by nature desire to know.[15]

Some people might wonder, though, whether many forms of "intelligence augmentation" really do lead to greater intelligence: Yes, pharmacological products designed to increase alertness and focus can give people a mental edge, but does the ability to store information in implanted computer chips or to get instant access to information via wireless systems that connect the brain to the Internet or databases have the perverse effect of making people dumber because they no longer think that they need to learn or even remember the material? After all, if this information is not actually in a person's memory, it cannot become dynamically integrated with other beliefs. This point was expressed well in Plato's *Phaedrus* (275a-b), where Socrates recounts what the Egyptian god Thamus said to the god Theuth shortly after Theuth invented writing:

> If men learn this [i.e., learn to write], it will implant forgetfulness in their souls; they will cease to exercise memory because they rely on that which is written, calling things to remembrance no longer from within themselves, but by means of external marks. What you have discovered is a recipe not for memory, but for reminder. And it is no true wisdom that you offer your disciples, but only its semblance, for by telling them of many things without teaching them you will make them seem to know much, while for the most part they know nothing, and as men filled, not with wisdom, but with the conceit of wisdom, they will be a burden to their fellows.

Plato is correct, to a point: A book does not understand its own words, whereas a mind can understand its beliefs. As Socrates says, knowledge "is written in the soul of the learner" and is not, as Phaedrus characterizes written language, a "dead discourse" that "may be called a kind of image" of living knowledge (*Phaedrus* 276a). Furthermore, Plato is right when he notes that many people do not develop their memories to the same extent as when they cannot rely on written sources. The ancient Greeks probably had noticed fewer people memorizing all or parts of the epic poems as literacy became more widespread. Plato also foresaw that some people would mistake possessing or superficially reading some abstruse text for understanding it. Nonetheless, Plato's pessimism about the written word (which, ironically, is recorded in one of Plato's *written* masterpieces) is falsified by history, for the invention of writing and later still of means of the mass dissemination of the written word have led to an otherwise unimaginable growth of human knowledge.

When Plato emphasizes the written word as a reminder, his model of writing is like that of a solitary person's grocery list, where the list functions like a person's memory about what provisions need to be purchased. Plato is correct that this sometimes is an important purpose of the written word. What scientist could remember all of the data she collects during experimental trials that stretch over months, and how could other researchers when hearing her results have confidence that her memory was not faulty? However, people often form *new* thoughts and have *new* understandings *because of* what they read. Sometimes this is as simple as looking up a fact (say, the year that the American Civil War ended) and then committing it to memory. Other times, one reads with a more critical approach that is like a dialog with the author:

> The kind of deep reading that a sequence of printed pages promotes is valuable not just for the knowledge we acquire from the author's words but for the intellectual vibrations those words set off within our own minds. In the quiet spaces opened up by the sustained, undistracted reading of a book, or by any other act of contemplation, for that matter, we make our own associations, draw our own inferences and analogies, foster our own ideas. Deep reading, as Maryanne Wolf argues, is indistinguishable from deep thinking. (Carr 2008)

Finally, because written works preserve ideas that can be widely circulated, these thoughts can spur widespread reflection, debate, and knowledge.

Contemporary concerns that future human thought will rely too heavily on computers strike me as similar to Plato's unease about our reliance upon written texts, so let us see how Plato's complaints would transfer to the digital age. First, will our memories become impoverished by relying on computers? There already is evidence to suggest an affirmative answer:

> This summer, neuroscientist Ian Robertson polled 3,000 people and found that the younger ones were less able than their elders to recall standard personal info. When Robertson asked his subjects to tell them a relative's birth date, 87 percent of respondents over age 50 could recite it, while less than 40 percent of those under 30 could do so. And when he asked them their own phone number, fully one-third of the youngsters drew a blank. They had to whip out their handsets to look it up. (Thompson 2007)

So, the objection may be warranted: One reason we use computational devices is to duplicate the "grocery list" aspect of writing and offload the task of remembering things like phone numbers and birthdays. Nonetheless, one could argue that taking trivial cognitive tasks of this sort and outsourcing them to computational devices frees up time and energy for more important and less rote mental tasks. We might squander the time-saving opportunities that technology creates for us, but technology creates space for us to focus on higher callings.

Furthermore, just as writing often either serves as an authoritative source so that the reader can form a corresponding belief (e.g., about when some historical event occurred) or provokes the reader to think in new ways, so the same will be true for electronic sources that will interface with the brain. There is no reason why "deep reading" experiences cannot be replicated as we received text, movies, or other experiences from chips and transponders directly connected to the central nervous system.

Finally, humans being so intimately wired to each other and to databases promise to accelerate the accumulation of knowledge and information. As in the aftermath of the invention of the printing press, the epistemological demands upon someone using electronic resources go up significantly, because the person must become savvy at determining which sources to consult and how to use those sources. The wide-open and ever increasing nature of the Internet means that everyone using the web for learning must become amateur epistemologists who can appreciate how to recognize trustworthy sources, calculate rough probabilities of truth, determine when a pre-existing belief should be discarded or revised in light of new "information," navigate conflicting claims on the web, and so forth. That is a good thing from the standpoint of virtue ethics, for these meta-cognitive skills are also among the most distinctively human. Consequently, these seemingly "unnatural" brain implants and the like allow us cultivate the more refined aspects of our human nature.

Every time the tools of intelligence augmentation advance, pedagogical questions arise about what an educated person should know or be able to do alone versus what an educated person should be able to discover from reputable sources or to do with the assistance of various tools. For instance, we expect children to learn addition and multiplication tables, but for mathematical problems beyond a certain level of complexity, we also expect children to learn how to use paper and pencil, which are tools of intellectual augmentation, to divide complex problems into manageable subroutines that are founded upon knowledge of the addition and multiplication tables and whose results are carried forward to subsequent subroutines (Clark 2003, 6, 74). After the invention of small and inexpensive calculators, controversy erupted over their place in the classroom, but eventually the ability to use these devices also was considered an important part of mathematical education. I expect more such pedagogical controversies in the future, and with the same results: Eventually educators will accept that tools of intelligence augmentation have a place in the curriculum—indeed, that the curriculum should instruct students how to use these tools properly.

Take the study of history as another example. If eventually we all will have implants that let us look up historical information easily, quickly, and at will, then what historical facts will a person need to carry around in the head? I am not suggesting that people will not need to learn the basic facts of history simply because they will be able to access this information easily—we should want people whose historical knowledge suffuses their understanding of themselves and the world, and someone almost entirely ignorant of history is unlikely to know what he or she is missing and should look up.[16] Nonetheless, it would be foolish to suggest that the teaching of history should go on as before with no recognition of how intelligence augmentation changes a person's lifelong epistemic condition. I suspect that educators will expect less memorization of facts and more understanding of when facts need to be investigated; less rote learning and more time devoted to higher-level tasks such as how to find credible primary and secondary sources, how to evaluate historical explanations, and so forth. In short, students will be pushed harder to acquire higher-order intellectual skills. Of course, this is a matter of degree: No self-respecting history teacher today would want students to do only rote learning with no analytical thought, but as intelligence augmentation becomes more widespread, the ratio of time and effort will tilt further toward analytical skills. How could a virtue ethicist see more of an emphasis upon critical reasoning as making us stupider or as alienating us from our humanity?

Moral Augmentation

In the previous section, I argued that intelligence augmentation can make us more intellectually virtuous or excellent. In this section I want to examine moral augmentation to achieve greater moral excellence, but first let me make a simplifying assumption. "Cognitive augmentation" was defined to exclude therapeutic treatments, but what would be a therapeutic treatment for moral virtues? I will assume that with the moral virtues, as long as a person was born with and did not lose unnaturally (e.g., lose from a brain injury rather than from disuse) the capacities that are needed to sustain moral character and behavior, the person would not benefit from therapeutic treatments but might benefit from moral augmentation. Thus, a person with bad moral character who could have had a good moral character if only he had been raised differently is not in need of a *therapeutic* treatment. By contrast, the sociopath lacking any ability from birth to be empathetic might benefit from, say, genetic therapy that addressed an underlying inability to be empathetic. (I use sociopaths as a hypothetical example: I have no idea whether genetic therapy could help cure them.) Of course, there are usually ranges of traits like empathy distributed through a population, and part of the explanation of that variability may be genetic, so it is difficult to say how unempathic someone must be before the person needs therapeutic treatment rather than moral augmentation. For our purposes, nothing hangs on drawing the line precisely, so I will assume that lines could be drawn somewhere to delineate normal ranges of capacities for empathy, moral reasoning, and other traits crucial for ethical character and deliberation.

The increasing sophistication of computers, robotics, and pharmaceuticals partly will drive developments in moral augmentation, but another factor is one that I mentioned in my discussion of the artificial retina, namely, the cracking of the mind/machine barrier. In discussing the example of an artificial retina, I focused on how the brain could receive signals from this artificial system, but there are already systems that allow the brain to directly control the operations of a mechanical device. Honda Motors has developed a robotic arm that, after a period of training/calibrating the software to properly interpret the person's brain activity, can be controlled directly by thought, and Toyota has developed a wheelchair that can be steered directly by thought.[17] Kevin Warwick even used the Internet to send signals from his nervous system in the U.S. to control a robotic hand in the U.K. (Warwick 2003, 135). Traditionally, the only thing over which someone's thoughts had direct control was that person's body, so what will it mean for our understanding of who we as embodied agents are if we, in some sense, extend our bodies to include artificial components under our direct control?

Philosophers and psychologists talk about a "body schema," which is how the mind, usually subconsciously, represents the body as something ready to be activated in motion, rather than as an object of perception.[18] For instance, I normally know of my hands' positions not, or not only, in the way that I consciously know of the location of, say, a painting on a wall but in a more intimate sense, namely, the typically background, subconscious understanding that my hands are in such-and-such places in relation to the rest of my body and are available for me to directly activate for various types of actions. A body schema is flexible, not only in the sense it must change as the body moves, but in the deeper sense that a body undergoes alterations that must be incorporated within the body schema. The brain must recalibrate the body schema as the body matures, as it receives injuries or contracts diseases that impair mobility, or as it loses a limb or other part of itself.

Technology also changes a body schema. When someone puts on a pair of sneakers, the body schema typically is reconfigured so that the shoes become part of that schema, and walking in them feels as natural as walking barefoot; one rarely thinks consciously about the shoes and about how one needs to lift one's feet higher to clear steps. For a person proficient in using them, a pair of stilts can be incorporated into the body schema as opposed to being something

represented as being outside the body and hence a tool to be consciously ma-
nipulated. Another example is sports equipment; a good baseball batter will
incorporate a bat into her body schema so that swinging the bat becomes like
swinging her arm. Perhaps most relevantly for the present discussion, there are
prosthetic limbs: A good prosthetic is comfortable, functional, and integrated
into the body schema so that the use of the prosthetic is easy and natural.[19]

In these cases of extended body schemas, the brain not only incorporates
something external into the body's integrated activities but also receives something that passes for somatic feedback: Using a cane, wearing shoes, swinging
a baseball bat, walking with a prosthetic leg—in all of these examples, the
brain receives feedback such as vibrations and changes in pressure through
sensory mechanisms that enable tactile perception to, in a peculiar way, extend
to the surfaces of the objects: When hiking in boots across rocky terrain, one
would say that one feels the sharp rocks that stick up from the ground and not
that one feels the boot poking into the foot. Similarly, if one is wearing thin
gloves and handling objects, then, in most contexts (except when, say, issues
of cleanliness or safety are involved), it would be odd to report that one didn't
touch the objects but only touched the inside of one's gloves. At the same time,
these external elements of the body schema are much duller organs of perception—one may not even feel tiny pebbles whose pressure is absorbed by the
sole of the shoe—and the perception is unaccompanied by pain or pleasure.
That is, when wearing a boot, one may feel pleasure or pain in the foot but not
in the boot. So, the body schema as a perceptual system outruns the body as a
site of pleasurable or painful sensations.

If for millennia simpler forms of technology have been transforming our
conception of the body, how much more of a transformation might occur from
the robotic arm or wheelchair that can be directly controlled by a person's
thoughts? I would guess that if something like Honda's robotic arm were attached to someone long enough as a souped-up prosthetic, it would become
part of the person's body schema and the person's brain would represent the
arm as a potentiality for agency much as the brain does for the limbs of the
body. Empirical research supports my hypothesis, for researchers have had a
61 percent success rate in getting monkeys whose arms are tied up to use their
minds to operate robotic arms to feed themselves: "The researchers said that the
movements were fluid and natural. The monkeys were able to use their brains to
continuously change the speed and direction of the arm and the gripper, suggesting that *the monkeys had come to regard the robotic arm as a part of their own
bodies* [my emphasis]."[20] Furthermore, the performance artist Stelarc has used
an attached robotic hand, attached to one arm and controlled by EMG signals
from abdominal and leg muscles, that can rotate and that can grasp, pinch, and
release objects: "By contracting the appropriate muscles you can activate the
desired mechanical hand motion. After many years of use in performances the

artist is able to operate the Third Hand intuitively and immediately, without effort *and not needing to consciously focus* [my emphasis]."[21]

Given the brain's ability to rewire itself as demands upon it change, I also would guess that if there were a physically separated, robotic device that could take mental commands and send back sensory information (i.e., if the brain and the device were connected wirelessly and had a mind/machine interface), then this robot also could become incorporated into a body schema so that the brain would represent the body as being in two locations simultaneously. We get a glimpse, however incomplete, of what multiple embodiment could be like from Kevin Warwick's experiment at the University of Reading's Department of Cybernetics with autonomous, mobile robots that use ultrasonic sensors: "In the experiment we fed the output from an ultrasonic sensor down onto my nervous system, rather than as an input to the robot. It was evident that my brain was able to make sense and good use of the signals that arrived." (Warwick 2003, 134) Unfortunately, Warwick does not describe how he made "good use of the signals" or what his conscious experiences were like. Perhaps multiple embodiment would be like daydreaming or dreaming while only half asleep, or being immersed in reading a novel that "transports" you to another place even as you maintain some awareness of your surroundings. In a daydream, one integrated set of conscious mental states, the daydream, comes to the fore while just above or below the threshold of consciousness there is an awareness of what is going on in the environment; the mind is consciously focusing on the daydream but remains alert to certain stimuli in the environment, which is why somebody can be snapped out of a daydream by hearing a loud noise. Multiple embodiment could be that way too, with one body getting the mind's conscious focus but the other body being monitored at a semi-conscious level. Harder to imagine is being equally consciously attentive to both bodies simultaneously—such a possibility raises questions about the meaning of the "unity of consciousness."

These robotic elements of our body schema would have fascinating implications for virtue ethics. There could be new and unforeseen sensorimotor virtues that involve the use of these robotic devices, as in Stelarc's performance art with the third hand, but I am more interested in what cognitive augmentation might mean for the moral virtues of self-control, regulation of appetites, and sociability. The moral virtues often relate to pleasures and pains, which are key motivators of human behavior. At present when we extend the body schema to include tools, the external objects can be loci of perception but not of pleasure and pain: If I held a wooden bat in my bare hands and hit a hard pitch, I would feel the ball hitting the bat, but if I felt any pain from the sharp impact, the pain would be in my hand, not in the bat. Suppose, though, someone could have an additional, artificial body (a "robo-body") that was separated from the biological body and that transmitted information to and received commands from the brain; what role would pleasures and pain play in that system? With unadulterated biological bodies, pleasure and pain are unavoidable aspects of experience. However, pleasure and pain could be optional for robo-bodies. If my robo-body bumped hard into something and suffered damage, one way of configuring the mind/machine interface would be for me merely to intellectually register the relevant facts, namely, that a collision occurred and caused damage to the robo-body. Another way of configuring the connection would be for me to feel pain from the collision. The distinction that I am describing, between receiving propositional information and receiving sensations from the robo-body, is like the difference between receiving "factual" linguistic reports about some situation and having lived experiences of that situation.

If I am correct that either type of linkage between brain and robotic device is possible, then which type would be preferable? Asking this question forces us to confront one of the hardest philosophical questions about conscious experience, which is, what are pleasure and pain (or, for that matter, any conscious experiences) necessary for? It might seem that, as long as I got sufficient information from the robo-body about what it was doing and how it was interacting with the environment, there would be no need for me, via the robo-body, to feel pleasures or pains, to have tactile sensations of warmth, to feel "hunger" for more energy as the battery pack ran low, and so forth. (Of course, even non-sensory reports from the robo-body might cause me to feel happiness or sadness, just as reading a sad magazine story can make me melancholic, but that would be an *indirect* effect on my overall state of mind.) Yet, the same sort of argument could be made for my biological body: As long as my brain can register that my leg is broken, why does the leg need to hurt? Pains and pleasures can have beneficial motivational force (e.g., an animal is motivated to avoid the pain it would feel when walking on a broken leg and hence does not worsen the leg's condition) that may help to ensure survival and reproduction for an animal. Indeed, all animals seem to experience sensations that are bound up with pleasures and pains, and this fact might imply that pains and pleasures have served an irreplaceable role in the survival of animals, for pleasures and pains are inherently motivational in ways that bare information is not. Even so, that cannot be the full story, because as robots and simpler mechanical systems demonstrate, a system that feels nothing at all can exhibit intelligent behavior. So, from an evolutionary perspective, the emergence of sensations, including those of bodily pleasure and pain, is mysterious.[22]

Even if we do not understand why this is so, sensations play a special role in enabling fast responses that bypass conscious but slower thought processes. For instance, a sudden pain can cause a quick reaction to minimize the pain, as when someone quickly jerks her hand from a hot surface. It would not immediately follow that robo-bodies must be loci of pleasures and pains for us—they are replaceable in ways that the animal's body was not in evolutionary history, so we could afford to have only slower, conscious responses that might occasion more robo-bodies to be damaged or destroyed, and some simple, self-preserving responses could be included in the robot's programming without the need for feedback from the brain; the biological analog would be reflex loops that bypass the brain.

Yet, given the way our central nervous system is currently wired, it seems likely that the best way to have fine motor control of a robotic device would be for these devices to give somatic feedback in the form of sensations. What would this mean for moral virtues that relate to pleasures and pain? Could we imagine a robo-body as a source of intemperate appetites and aversions that lead to an overindulgence of short-term and superficial pleasures or, like the cowardly person, avoidance of necessary pain? To take an extreme example, a therapeutic implant to stimulate women's spinal nerves to give them an orgasm at the push of a button has been developed for clinical trials (Sample 2001, 23)—if we could configure an artificial body that could engage in sex and cause the same sort of orgasmic response, then the ethical questions that we associate with the proper role of sensuality and intimacy in a full and rewarding life will carry forward into the world of artificial bodies.

Nonetheless, what is striking in these scenarios is how easily some behavioral ideals with regard to the robo-body could be achieved by reprogramming the robo-body's sensory responses to the environment. Regulating the robo-bodies' pleasure and pain responses would not have to be an authentic struggle of the soul but instead could be a programming challenge to be solved by experts in computer science—if someone was becoming addicted to using his or her robo-body to have sex, then have the engineers turn down the orgasmic response mechanism. Alternatively, technology might put these tools for regulating the bodily pleasures and pains in the user's hands—if something in the robo-body hurts more than it would for a temperate and virtuous person, recalibrate the settings to produce lower pain intensity in these cases. If something does not give as much pleasure as it would for the temperate and virtuous person, recalibrate in the opposite direction. We then would find that moral problems related to bodily (or, more accurately, robo-bodily) desires switch from being first-order to second-order problems. Let me explain.

On the Aristotelian picture of virtue ethics, the ideal life belongs to the happy person whose soul exhibits a harmony of elements that aim at what is good. This person had to have the right sort of upbringing to form an appropriate second nature or *character* so that what is pleasant or painful to that person will reinforce the intellectual judgments that such a person correctly makes about how to live. *Because we desire to have pleasure and to avoid pain, our character, by determining within the limits of human nature what, when, and to what extent we find something pleasurable or painful, also determines to some extent what our first-order desires are.* Hence, not only will the virtuous person have first-order desires for good things, but when thinking about those first-order desires, the person will have a second-order desire for the first-order desires to remain as they are. By contrast, a continent person may do the right thing, but only by overcoming intemperate first-order desires. This person would prefer not to have disharmony in the soul and consequently would normally have a second-order desire for better first-order desires.[23] Yet, because the first-order desires are a product of the person's past training and upbringing, the continent person's second-order desire to change the first-order desires is impotent.

Now consider someone with a robo-body. Programming a robo-body to change when and how it feels pleasure or pain would be like consciously controlling the robo-body's first-order desires instead of being stuck with first-order desires that were determined by the person's upbringing and moral education. Because the robo-bodied person would have control over that body's first-order desires, the second-order desires would not be impotent. In traditional moral development, modeling, reinforcing, and punishing behaviors and teaching principles of moral reasoning shape a person's character; someone is born neither good nor bad but is made so by the environment. By contrast, a robo-body would be a blank slate that the *agent* could write upon. Consequently, unlike the continent person in Aristotle's philosophy, a person with intemperate desires rooted in how his robo-body gave him pleasure or pain could not place the blame on other people for not having raised him correctly, and the person with intemperate sensuality related to the robo-body would be more culpable for bad character, unless determinism is so complete that the person's upbringing also effected what her *second*-order desires would be.

The ability to program an artificial body's pleasure and pain responses might diminish the respect we have for virtues related to those pleasures and pains. From an Aristotelian perspective, one reason to admire the virtuous person is not only that such a person embodies what is best in humanity and helps others but also that becoming such a person is an *accomplishment*. Because nobody is born good or evil, it is through striving for goodness, with proper tutelage and reinforcement, that one becomes a good person. Like a great artist or athlete, a morally virtuous person is the superlative outcome of a difficult process of training and education, and that makes this person's character all the more awe inspiring. Contrast that model with someone with a robo-body to supplement (or perhaps one day to replace) his biological body. His pleasure and pain responses for the artificial body would be the product of how the robo-body's outputs were coded as pleasurable or painful (or the magnitudes of these pleasures and pains), and this no longer would seem like an admirable accomplishment by the virtuous person.[24] Perhaps the only thing to praise as a moral accomplishment by this person would be that the person had the right second-order desires. So, the person would get credit for wanting to be a good person, but much less credit than people now receive for having those virtues. Would that change the credit we would give someone for *actions* that flow from his or her virtuous character that was purchased so cheaply? Would moral virtue cease to be a source of admiration when acquiring it would be less like undergoing a rigorous training program and more like downloading software? Here a second-order evaluative question arises that Kantian ethics, with its emphasis upon free will, tends to downplay: The evaluation not simply of the will's choice but of how that person's will came to be the sort of thing that would make *that* choice. I will return to these questions shortly.

Let us now turn out attention to future uses of brain implants and drugs for moral augmentation. Just as anti-depressant medications allow some patients to bypass talk therapy and get remedies from bottles of pills, which makes the person seem less like an *agent* in his own recovery and more like a *patient* who is cured by the efforts of others, so we may be approaching the day when medications and brain implants could help regulate a person's pleasures, pains, and appetites. The brain is a chemical and electrical system, although an extremely complex one, that is subject to modification and redesign like any other physical system, so there is no reason why someone could not have drugs or implants to make the person's affective responses closer to that of the wise and just person. These recalibrations could be done in a variety of ways: procedures to activate or suppress the expression of genes,[25] drugs to change the brain's chemistry, implants to enhance or regulate the brain's electrical activity, nanotechnology that could be injected into the blood stream to help regulate the brain's chemistry and tweak a person's character, and so forth. The challenges of eliminating unwanted side effects from these procedures would be large, and any steps in this direction would occur slowly as we learned more about the central nervous system, the consequences of experimental procedures, and so on. Nonetheless, decades of progress in medicine and technology suggest that the trend will be to understand the central nervous system as a physical thing that can be supplemented or re-engineered to change a person's character.

Aristotle thought of character as an acquired but enduring state—a permanent reconfiguration of someone's inner constitution and mental life that tends to produce certain types of affective responses and behaviors in certain types of situations. If the sorts of moral augmentations that I have suggested are on the horizon for humanity, then our notion of character would have to change. First, if I have to rely on drugs, implants, nanotechnology, or any other artificial resource to maintain a certain type of moral character that causes me to take pleasure in morally praiseworthy things and feel pained by morally shameful things, then my character is a state not of an unadulterated biological person but of what I become with certain types of moral augmentation. Take away those enhancements, and you also take away part of my character. Does that mean that I never had that character to begin with, because it was not the sort of enduring internal state that Aristotle describes? No, because the idea that there is a "real" me that can be separated from culture, technology, and environment is a fiction. Isolate a normal and sociable person, and that person may become unstable and anti-social, but that does not mean that mean that the person was "really" anti-social all along. Similarly, deprive someone of food, and even a mild-mannered person may become grouchy and potentially violent, but that does not show that the true person had some underlying, violent disposition. Although some people may worry about where they end and the technology begins if we begin taking drugs and using implants to help regulate our characters; that is to buy into a false picture of the human condition: All of us are culturally and technologically embedded and dependent creatures. To think that the "real" persons are the imaginary versions of ourselves who live outside these social and technological networks is not true to human history, never mind to the increasingly "enhanced" humans of the future.

Nonetheless, some readers who feel uneasy about technologically assisted virtue might have two further questions. First, is there not a special type of value and dignity that comes from having to struggle against unruly appetites and desires? If this line of thought is taken to its most extreme form, we might accept Peter Abelard's rejection of the Platonic and Aristotelian ideal of a harmonious soul and instead praise the existence of unruly passions that we must overcome to be good. Abelard's point is not the consequentialist idea that by facing adversity, one becomes the sort of person more likely to succeed in future struggles. Rather, Abelard believes that holiness of the will is enhanced by overcoming evil temptations, because the will's very purpose is to conquer obstinate desires and appetites. According to Abelard, the will's struggle against fallen human nature has intrinsic, not merely instrumental, value.[26]

I side with Plato and Aristotle against Abelard. Abelard's theory conflates the admirable and the preferable. Although we would greatly *admire* someone who survived a natural disaster and held onto life against great odds while experiencing unfathomable suffering, we (or the person in question) should not *prefer* that this person undergo those experiences. It would be morally perverse for us to wish a tragedy upon someone so that we could delight in the person's perseverance and grit, however admirable those traits may be. By parity of reasoning, even if Abelard is correct that we might have the utmost respect for someone who always did the right thing, but only by strenuously and painfully wrestling against her wicked sensible nature, we (or the person in question) should not prefer that the person's passions be at odds with reason. So, Aristotle is correct: The virtuous person is to be preferred to the continent person, and consequently, technologies that help us be more virtuous have moral value.

A second worry, which I mentioned earlier in relation to robo-bodies, is not *whether* these virtues associated with having the proper appetites are desirable but the ethics of *how* those virtues are acquired. Even people who accept that the temperate character is to be preferred to the continent one might think that a virtue purchased too inexpensively is thereby cheapened. Traditionally, a good character has been something like a work of art, not a commodity—that is, it has been the outcome of a humane, personalized, and non-commercial process, not a product to be purchased like a pair of sneakers. Would something important be lost if someone's character were assembled like a computer from off-the-shelf components rather than cultivated authentically and strenuously under the guidance of wise mentors?[27]

This worry is Abelard's fallacy in a different guise, and my response is similar. Yes, any person who worked hard to acquire a good character is going to be *admired* more for successfully struggling to achieve a noble but difficult end. Yet, we should not *prefer* that people struggle in that way. Perhaps, though, someone who had to work hard to acquire particular virtues will, in the process of doing so, acquire other important virtues like perseverance, resilience, and so forth, whereas if the good things in life come too easily to somebody, that person will lack those virtues. That is a legitimate concern, at least as long as we cannot artificially induce perseverance, resilience, and so forth. Even so, moral augmentation sometimes would be justified, given that our current ways of helping people acquire the moral virtues of properly tuned appetites are unreliable. Many, if not most, people have unruly appetites that lead them astray or, should they try to regulate them at all, are a source of internal struggle and unhappiness. The popular demand for motivational speakers, diet and exercise gurus, and addiction counselors is evidence that many people find it hard and unnatural to do the right thing. To what extent these problems are due to nature, nurture, or both is hard to say, although rising levels of obesity in the advanced and developing economies strongly indicate that culture and economics are significant causes, at least in this case. Whatever the explanations are, there appears to be no easy fix, either at the individual level (e.g., personal resolutions to go on a diet, exercise more, spend less, and so forth) or, without a violation of a free peoples' sense of liberty, at the social level. Consequently, a technological "safety net" that artificially induces virtues is eminently reasonable. Although parents, teachers, and other role models should bear in mind the meta-virtuous benefits that come from children acquiring appetitive virtues the old-fashioned way, we must be realistic and assist people who, for whatever reason, did not acquire these virtues as youth and do not seem to be able to do so through their own efforts. As a last resort, moral augmentation could be analogous to bariatric surgery to treat obesity when traditional weight-loss techniques have failed. Even people with mostly healthy appetites might find benefit from further, artificial tuning.

So, subject to the caveats just mentioned, future people will be justified in taking an artificial path to Aristotelian temperance and virtue. Religious or moral convictions will lead some people to forego artificial virtue enhancers, just as some people willingly take vows of poverty or choose, like the Amish, to abstain from modern conveniences. Most people, though, will accept the opportunity to become better and happier people—better because their sensible natures would lead them in the direction of doing the right thing, and happier (at least compared to continent people) because doing the right thing would be at odds less often with their sensible nature. Although how they acquire virtue initially might have some stigma to it as a sign of moral weakness (again, the comparison to bariatric surgery is instructive), over time this should disappear as it becomes more common.

So far I mostly have discussed the virtues that are related to what sorts of appetites or, more broadly, emotional and appetitive reactions one has to various circumstances, but what about prudence, the practical wisdom to rationally make difficult moral decisions, to override one's passions when necessary, to weigh competing goods, to choose a worthy and just path in life, and so forth? Here too augmented cognition may assist us. First, some political and ethical decisions are hard to make because relevant consequences can be difficult to anticipate. By collecting and analyzing large amounts of data, computers already are helping reduce this uncertainty in areas of our lives such as global warming. Computer modeling can predict outcomes that never would have been foreseen by unaided humans because of the immense number and complexity of variables, and simulations can help people make better decisions. Software coaches or assistants also will play an increasingly important role in our lives. Imagine having software collecting data about you and your performance at various tasks and then using those data to help you make realistic plans for completing a major project. Imagine too if this program helped to monitor your progress, nudge you to stay on task, and motivate you not to procrastinate by showing what new timetable would be required were you to fall behind schedule. None of these programs is far removed from the realm of possibility, and without reducing your agency or freedom, they would encourage more prudential decision-making.

Coda in a Minor Key

I began this essay by asking whether cognitive augmentation, which I defined as an increase in mental powers that results from using one or more improvised, non-therapeutic tools, alienates us from our humanity. That question is best answered from the perspective of virtue ethics, and I argued that we are by nature problem solvers and builders and users of tools that transform our external and internal or bodily environments. Hence, tools that further our natural ends of wisdom and morality do not estrange us from our humanity but put us in closer contact with it.

Nonetheless, I am not blindly optimistic about what the future may bring. Some difficult questions will need to be confronted about access to these tools.

Is there a right to cognitive augmentation, like there is to education (Warwick 2003, 136)? Will the government need to guarantee that all citizens have access to these tools because they help realize human nature and because people without these tools would be at a disadvantage? A worst-case scenario is that people who choose to undergo intelligence augmentation are going to have an edge over the rest of us, and if these tools are distributed through the free market, a vicious spiral of social inequality could ensue where the rich could afford to make themselves smarter, which let them get richer, and so on (Crittenden 2002, 140). Many nations, on the basis of moral and economic considerations, have decided that they are obligated to offer free, universal education to their citizens, and I hope that they will decide to do the same with some tools of intelligence augmentation. In America, though, the omens for this sort of equality of opportunity are not good: Many poor families cannot afford computers or high-speed Internet access, and the latter is unavailable in some regions of the country. Furthermore, even if all people in the advanced economies were to get access to the tools of intelligence augmentation, what about people in poor countries? Will global inequality increase?

These tools of intelligence augmentation also might create new vulnerabilities. Hackers already break into secured computer accounts and steal valuable data, but how much more vulnerable will people be when it is possible to hack into the brain's implants? Furthermore, the opportunity for governments to monitor citizens will be even greater if governments can access devices whose memories can be read like any other computer (Crittenden 2002, 141).

Another frightening possibility is that this technology will commoditize us and make us too alike, that the idiosyncratic biological elements that give each person his or her individuality will be overwhelmed by the installation of standardized hardware and software programs and the ingestion of drugs designed to optimize intelligent and ethical behavior. Perhaps as we get better at using tools to augment elements of our universal human nature, we run the risk of eliminating the quirks that make concrete individuals who are more than their universal human essence interesting and creative.

The most frightening scenario of all might be if our tools of intelligence augmentation become smarter than we are and develop their own agency. What, then, would prevent these machines from turning against us? Fortunately, computer scientists are aware of the risks of runaway artificial intelligence and are discussing, for instance, whether advanced artificial-intelligence research needs to be conducted in high-security laboratories.[28]

So, the future holds potential opportunities and perils, as always is the case with powerful new technologies. More people of all backgrounds need to begin looking into the future to anticipate and head off possible dangers and to push society toward using these tools to help everyone live a fuller, more human life.[29]

Notes

1. I would like to thank Professors Severin Kitanov (Philosophy), Keja Valens (English), and Patricia Johnston (Art), members of my reading circle of Salem State College's 2009 Research and Writing Initiative (RWI), for their feedback on two earlier drafts of this paper and for encouraging me to think that there might be some value to these speculative musings. Thanks also must go to Salem State College for its support of the RWI.

2. I will touch on this topic briefly at the end of the paper, but for now I am putting aside the persistent worries of dystopian science fiction novels and movies such as *Frankenstein, I Robot,* and *2001: A Space Odyssey* in which the created object has its own agency and turns against its creator. The primary aim of this essay is to examine the concern that technology may corrupts us from within, not that it may enslave us from without.

3. Although in this paper I write as if I were a virtue ethicist, that is not entirely accurate. I consider myself to be a moral pluralist and think of virtue ethics as one element, along with aspects of deontological and consequentialist theories, of a more comprehensive ethical theory. Virtue ethics is particularly illuminating in exploring human nature and the contours of a good life for humans, which are the topics that I want to pursue here in relation to cognitive augmentation.

4. Ruth Garrett Millikan (1984) introduced the term "proper function," as well as the theoretical framework in which it is embedded.

5. The line between doing something entirely new and extending what we already do is unclear: Do aircraft extend our powers of locomotion or give us a new power of flight? The answer must be "both." Fortunately, this distinction will not be particularly important for our purposes.

6. Some writers are enthusiasts for transhumanism, the theory that "humans are a ***trans***itional form of a species intent on ***trans***cending itself—we are a species on the way to a new and yet unimaginable posthuman form of intelligent life" (Doede 2009, 34). Transhumanism raises a host of important questions, but what I am exploring in this essay is how we cognitive augmentation can supplement our biological resources with technology to further develop our human moral and intellectual virtues. Developing technology of this sort may also move us closer to the day when we could dispense with biological entities altogether, but whether that would be a good thing is not a question that I have space to explore in this essay. The influential arguments first presented in Searle (1980) that computers lack original intentionality and hence do not understand anything have to be answered before we could confidently say that humans could "upload" their minds to computers and live on as cybernetic entities. Searle famously argues that consciousness and original intentionality arise from inherent properties of the brain's biological constitution, properties that are not duplicated in silicon chips. If Searle's arguments can be refuted, it would turn out to be a paradox that the tools of cognitive augmentation that made us more human also put us closer to the day when we could be superseded by or transformed into machines.

7. See also Gregory M. Lamb, "Strange Food for Thought," *Christian Science Monitor*, 17 June 2004, http://www.csmonitor.com/2004/0617/p14s01-stct.html.

8. Standard eyeglasses, hearing aids, and so forth are therapeutic devices and hence are not examples of intelligence augmentation, but souped-up versions of the same would be devices for intelligence augmentation.

9. The beauty of Moore's Law, which states that semiconductor capacity will double roughly every eighteen to twenty-four months and hence that the price of semiconductors should halve in about that same time period, is that the hypothetical six million dollar price would drop steadily year after year, so that we could have implants for the masses and not just for the rich. Moore's Law has been accurate for decades, but there are no guarantees that it will be so in the future. For a gloomy take on the possibility that Moore's Law soon will be falsified, see Saul Hansell, "Counting Down to the End of Moore's Law," *New York*

Times, 22 May 2009, http://bits.blogs.nytimes.com/2009/05/22/counting-down-to-the-end-of-moores-law/.

10. Cascio (2009, 96) argues that we are unlikely to have computer chip implants because they would be unstable: "Would you want a chip in your head made by the same folks that made your cell phone, or your PC?" Cascio (2009, 96) also thinks that these implants would become obsolete too quickly: "Who would want a 2025-vintage brain when you're competing against hotshots with Model 2026?" Let me address each of Cascio's concerns separately. First, stability does not have to be an impediment to having these implants: We have computer chips guiding aircraft, controlling automobile emergency steering systems, operating life-support systems at hospitals, monitoring nuclear reactors, directing missiles and warheads, and so forth. If we can entrust our lives to computers for these purposes, then we can rely upon implants systems with suitable redundancies to provide extra memory and processing power. Secondly, if frequent major surgeries were necessary to replace obsolescent or broken implants, then we might decide that they were not worth the trouble, but, although this may sound disturbing today the implants probably could be designed so that the they would be in a compartment easily accessible to a trained specialist—almost like battery compartment.

11. Leslie Berlin, "Kicking Reality up a Notch," *New York Times*, 12 July 2009, p. BU3, http://www.nytimes.com/2009/07/12/business/12proto.html,

12. Germ line and somatic gene therapies introduce many complicated moral issues that I cannot address in this essay. I will note here only that germ line genetic engineering in a single generation has the potential to alter the gene pool for subsequent generations and introduce persistent changes to the character of the species.

13. Steve Mizrach, "Should There be a Limit Placed on the Integration of Humans and Computers and Electronic Technology?" http://www.fiu.edu/~mizrachs/cyborg-ethics.html.

14. Perhaps, though, these future developments will happen so gradually that most people never have a moment when they think that these changes are too odd or unnatural to accept (Cascio 2009, 100). For instance, augmented reality is likely to begin with external devices such as eyeglasses or contact lens—by the time implants are available to integrate information into our visual fields, augmented reality might be such a vital part of our lives that we would be grateful to internalize and not have to fuss with the device.

15. That is not to say that augmented intelligence would have no downsides, even after excluding the inevitable unintended and unforeseen medical consequences that will arise for some people as these devices and procedures are refined. For instance, as of 31 August 2009, Google's free translation service, Google Translator (http://translate.google.com/?hl=en#), lets people translate web pages or documents from any of fifty-one languages into another of those languages, which is an astounding form of intelligence augmentation, and there are numerous speech-to-speech translation programs also available. Google Translator is far from perfect: From poking around with it, I would tentatively conclude that it translates from Germanic or Romance languages into English well but can produce garbled English sentences when the source language is, say, Asian. Nonetheless, the software will steadily improve (and I am sure that the security establishment of the United States already has far more sophisticated and accurate programs at work for intercepts of e-mails, phone calls, letters, and so forth), and even if these programs never have the reliability or finesse of a knowledgeable human translator with a literary ear, it is striking that we already are at a point when a child with a bit of tech savvy now has greater powers of translation than even the most accomplished polyglots of the past, for how many people could translate into and out of fifty-one languages, even if some of the translations are fairly primitive? Speech-to-speech translation programs also have made great strides, and if they eventually become integrated into a system of augmented reality, we might be able to "hear" roughly simultaneous translations of the spoken word as if a human translator were working by our side.

Translation systems promise to tear down global barriers of communication, but technology often has potential downsides, namely, in this case, that fewer people might bother to learn a second (or third) language. When one acquires a foreign language, one learns a different way of conceptualizing the world, and there are ideas, puns, distinctions, ambiguities, and so forth that elude translation. Will the simplicity and ubiquity of translation devices mean that fewer of us will want to do the hard work of studying other languages? Will many people become more epistemically provincial by not knowing how to conceptualize the world like members of another linguistic community? These are legitimate worries, but nothing in the nature of the technology would prevent anyone from learning other languages. To not learn a second language would be a *choice*, a *bad* choice, and it would be irresponsible to blame technology for our own choices.

Another potential problem is that if a particular translation program for our implants achieved a monopoly position (as Microsoft Windows has done with operating systems for personal computers), the program's inevitable flaws or debatable translation choices would systematically warp communication between linguistic groups. This sort of problem does not arise when we have computers take on cognitive tasks for which there is an unambiguously correct answer, as in the value of a mathematical equation, but when it comes to tasks where even expert opinions might differ, technological homogeneity is not a good thing.

16. The sorts of problems that educators already face when letting students use laptops during "closed book" exams will be magnified with implants: If educators want to ensure that students learn certain basic facts of history, then during an exam teachers would need a way to disable or disconnect a student's implanted devices that store information or that access the Internet.

17. Yuri Kageyama (Associated Press), "Honda Connects Brain with Robotics: Honda Shows New Technology Linking Brain Thoughts with Robotics," ABC News, 31 March 2009, http://abcnews.go.com/Technology/wireStory?id=7215636. Associated Press, "Toyota Technology Has Brain Waves Move Wheelchair," *Boston Globe*, 29 June 2009, http://www.boston.com/business/technology/articles/2009/06/29/toyota_technology_has_brain_waves_move_wheelchair/.

18. My discussion of the body schema is indebted to Noë (2009, 77-80). Clark (2003, 59-62) has an interesting discussion of the brain's ability to recalibrate the "body image," which is "our sense of our own bodily limits and bodily presence" (Clark 2003, 59). The phenomena that he explores is somewhat different, insofar as the experimental results he mentions involve fooling the brain into thinking that it is experiencing sensations where it is not actually receiving stimuli (e.g., feeling as if one's nose which is being tapped extends two feet out from one's face). These results reinforce the idea that the brain has a flexible way of subconsciously thinking about what constitutes the body: "*despite* the probable presence of some preset genetic components in our body-images, there is also—and simultaneously—large scope for continual revision" (Clark 2003, 61).

19. Of course, anything that is in a body schema can "objectified" in experience too. For instance, if I use tweezers to remove a splinter from my foot, my foot is part of the body schema and hence is cognized as something potentially to be activated or moved but also is represented as an object of potential action. The same can be true for tools that are incorporated in the body schema.

20. BBC News, "Monkey's Brain Controls Robot Arm," 28 May 2008, http://news.bbc.co.uk/2/hi/science/nature/7423184.stm. Remarkable video of a monkey controlling the robotic arm is available on that page and other web sites. More examples of humans' and other animals' minds directly controlling devices can be found in Warwick (2003, 133).

21. From "The Involuntary, the Alien & the Automated: Choreographing Bodies, Robots & Phantoms" on Stelarc's web site, http://www.stelarc.va.com.au/articles/index.html. Video of Stelarc using the third hand is available on YouTube and other web sites. Clark (2003, 115-119) has a perceptive discussion of the philosophical implications of Stelarc's art.

22. At bottom, this is the same problem as the problem of other minds, the "zombie" problem, and other philosophical puzzles rooted in the logical possibility that intelligent human behavior could be linked to different or even to no subjective conscious states.

23. It is possible, though, for the continent person to have an irrational third-order desire for a different, irrational second-order desire for her actual, irrational first-order desires. In this case, the rational second-order desire may be stronger than the irrational third-order desire, so the person still may do the right thing most of the time, even though doing so is unpleasant.

24. Here and elsewhere, I pretend that the body entirely determines the nature of bodily pains or pleasures. In truth, although the stimuli from the body sets a range of conscious experiential responses, the person's state of mind will determine where in that range the experience falls. Thus, nobody can find someone breaking his leg to be pleasant, but a person's attitude or strategy of stoicism, resignation, optimism, deflected attention, depression, and so forth can influence how badly it hurts. Throughout the essay, let this proviso be implicit for my discussions of bodily pleasures and pains.

25. The regulation of genes, which does not directly alter the gene pool, is to be distinguished from the artificial selection of genes. The latter changes the gene pool and consequently, for reasons that I explained earlier, would not count as a form of cognitive augmentation.

26. "What if this [lustful] willing is curbed by the virtue of moderation but not extinguished, stays for the fight, holds out for the struggle, and doesn't give up even when defeated? For where is the fight if the material for the fight is absent? Where does the great reward come from if there is nothing serious we put up with? When the struggle has passed, there is no fighting left but only the receiving of the reward. We struggle by fighting here in order that, triumphant in the struggle, we might receive a crown elsewhere. But to have a fight it's proper to have an enemy who resists, not one who gives up altogether. Now this enemy is our bad will, the one we triumph over when we subject it to the divine will. But we don't entirely extinguish it, so that we always have a will we might strive against." (Abelard's *Ethics* 1.22). I don't know whether Kant read or was familiar with Abelard's writings on morality, but Abelard's ethics has precursors of Kantian ethics. Kant, though, is equivocal about whether a will that acts from duty and against inclination is morally superior to a will that acts from duty and in accordance with inclination. For a discussion of this topic in Kant's moral philosophy, see Wike (1994, 39-48).

27. Although we are examining this question in the context of moral augmentation, similar questions (and answers) would arise in relation to intellectual augmentation.

28. John Markoff, "Scientists Worry That Machines Might Outsmart Man," *New York Times*, 26 July 2009, p. A1, http://www.nytimes.com/2009/07/26/science/26robot.html.

29. This essay is dedicated to the memory of Louis Pojman— a wise philosopher, caring mentor, and good friend.

References

Abelard, Peter. 1995. *Ethics*. In Ethical writings: His "Ethics or 'Know yourself'" and his "Dialogue between a Philosopher, a Jew, and a Christian," trans. Paul Vincent Spade. Indianapolis: Hackett.

Adams, Frederick A. and Kenneth Aizawa. 2001. The bounds of cognition. *Philosophical Psychology* 14 (1): 43-64.

---. 2008. *The Bounds of Cognition*. Malden, MA: Blackwell.

California Institute of Technology. 2009. Artificial-retina project designed to restore sight to the blind. *Science Daily*, 4 August. http://www.sciencedaily.com/releases/2009/08/090804132810.htm.

Carr, Nicholas. 2008. "Is Google Making Us Stupid?" *Atlantic*, July/August. http://www.theatlantic.com/doc/200807/google.

Cascio, Jamais. 2009. Get smart. *Atlantic*, July/August: 94-100.

Clark, Andy. 2000. "The Extended Mind." *Analysis* 58 (1): 7-19.

---. 2003. *Natural Born Cyborgs: Mind, Technologies, and the Future of Human Intelligence.* Oxford: Oxford University Press

---. 2005. "Intrinsic Content, Active Memory and the Extended Mind." *Analysis* 65 (1): 1-11.

---. 2006. "Material Symbols." *Philosophical Psychology* 19 (3): 291-307.

--- and David Chalmers. "The Extended Mind." *Analysis* 58 (1): 7-19.

Crittenden, Chris. 2002. "Self-Deselection: Technopsychotic Annihilation via Cyborg. *Ethics & the Environment* 7 (2): 127-152.

Cybersenses. 2009. *Scientific American Frontiers* #1509. PBS. http://www.iptv.org/video/detail.cfm/4167/samf_20090614_cybersenses.

Doede, R. P. 2009. "Polanyi in the Face of Transhumanism." *Tradition and Discovery: The Polanyi Society Periodical* 35 (1): 33-45. http://www.missouriwestern.edu/orgs/polanyi/TAD%20WEB%20ARCHIVE/TAD35-1/TAD35-1-fnl-pg33-45-pdf.pdf.

Garreau, Joel. 2004. *Radical Evolution: The Promise and Peril of Enhancing Our Minds, Our Bodies—And What It Means to be Human.* New York: Doubleday.

Menary, Richard. 2006. "Attacking the Bounds of Cognition." *Philosophical Psychology* 19 (3): 329-344.

Millikan, Ruth Garrett. 1984. *Language, Thought, and Other Biological Categories: New Foundations for Realism.* Cambridge, MA: A Bradford Book, MIT Press.

Noë, Alva. 2009. *Out of Our Heads: Why You Are Not Your Brain, and Other Lessons from the Biology of Consciousness.* New York: Farrar, Straus and Giroux, Hill and Wang.

Plato. 1952. *Plato's Phaedrus.* Translated by R. Hackforth. New York. Reprinted in *The Collected Dialogues of Plato, Including the Letters.* Edited by Edith Hamilton and Huntington Cairns. Includes corrections from the second printing. Bollingen Series LXXI. Princeton: Princeton University Press, 1963.

Rindos, David. 1985. "Darwinian Selection, Symbolic Variation, and the Evolution of Culture." *Current Anthropology* 26 (1): 65-77.

Rupert, Robert D. 2004. "Challenges to the Hypothesis of Extended Cognition." *Journal of Philosophy* 101 (8): 389-428.

Sample, Ian. 2001. "Push My Button: Electronic Implants May Help Women Who Can't Orgasm Any Other Way." *New Scientist*, Feb. 10: 23. Expanded Academic ASAP. Gale. Salem State College. 12 Sept. 2009 http://corvette.salemstate.edu:2350/gtx/start.do?prodId=EAIM.

Searle, John. 1980. "Minds, Brains, and Programs." *Behavioral and Brain Sciences* 3 (3): 417-457.

Selinger, Evan and Timothy Engström. 2007. "On Naturally Embodied Cyborgs: Identities, Metaphors, and Models." *Janus Head* 9 (2): 553-584. http://www.janushead.org/9-2/SelingerEngstrom.pdf.

Sterelny, Kim. 2004. "Externalism, Epistemic Artefacts and the Extended Mind." In *The Externalist Challenge*, ed. Richard Schantz, Current Issues in Theoretical Philosophy, 239-254. Berlin: Walter de Gruyter.

Talbot, Margaret. 2009. "Brain Gain: The Underground World of "Neuroenhancing" Drugs." *New Yorker*, April 27. http://www.newyorker.com/reporting/2009/04/27/090427fa_fact_talbot.

Thompson, Clive. 2007. "Your Outboard Brain Knows All." *Wired*, 25 Sept., http://www.wired.com/techbiz/people/magazine/15-10/st_thompson.

Warwick, Kevin. 2003. "Cyborg Morals, Cyborg Values, Cyborg Ethics." *Ethics and Information Technology*, 5 (3): 131-137.

Weiskopf, Daniel A. 2008. "Patrolling the Mind's Boundaries." *Erkenn* 68: 265-276.

Wike, Victoria S. 1994. *Kant on Happiness in Ethics.* SUNY Series in Ethical Theory. Albany: State University of New York Press.

Natural Rightism and the Biogenetic Debate

Grant Havers

The emergence of biotechnology in the twentieth century has not only transformed our attitude towards the mission of the medical profession. While advances in genetics, assisted reproduction, and regenerative medicine have encouraged our hopes for longer and healthier lives, technological change has also altered our understanding of human nature and ethics. If biotechnology can easily manipulate and even redefine what exactly a human being is, it must follow that our concept of what is morally owed to human beings and even humanity as a whole changes as well.

The ethic of utility often dominates the debate over the degree to which societies should permit the advance of biotechnology. The apparent practicality of calculating the pains and pleasures that may derive from biogenetic experimentation fits well with our overall valuation of science in our culture. For example, if the benefits of genetically producing "designer babies" with higher intelligence and more stable emotions outweigh the costs, then utilitarians are likely to support this hypothetical experimentation. Yet not all ethicists are convinced that the ethic of utility is the best or only form of moral reasoning that scientists and governments should apply to the debate over biogenetics. Since utilitarians are mainly concerned with what can be calculated as a cost or benefit (pain or pleasure), they are likely to be uninterested in abstract and unquantifiable goods like human dignity and the sanctity of life.

The school of *natural rights* is morally committed to the preservation of these goods. For this reason, "natural rightists" (as I shall call them) take aim at utilitarian reasoning as a defective approach to the value and meaning of human life. Indeed, the defenders of natural rights argue that their ethic is the *only* mode of moral reasoning that can resist the utilitarian temptation to reduce all ethical questions to a calculation of changeable interests. As one prominent natural rightist, Francis Fukuyama, explains, we must believe that all human beings are endowed with rights by nature in order to resist the most extreme forms of economic calculation of the value of life. In the simplest terms, natural rightists worry that the changeability of utilitarian calculation of interests renders any stable commitment to human dignity weak and ineffective: "Rights trump interests because they are endowed with greater moral significance. Interests are fungible and can be traded off against one another in a marketplace; rights, while seldom absolute, are less flexible because it is hard to assign them an economic value."[1] The economic rationale behind experimentation in "designer babies" does not, for example, justify the surrender of basic human dignity and the sanctity of life to the marketplace. A natural rights ethic requires human beings to prevent the market of desires and preferences from dictating the value of life.

Natural rightists are convinced that the right to a life lived with dignity preempts any utilitarian justification for tampering with this right through technology. As Fukuyama puts it, there is a "stable human essence with which we are endowed by nature,"[2] which is the foundation for natural rights. In other words, *nature* rather than the market must set the limits for intervention into the genetic realm. While natural rightists do not object to the need to employ medical means to fight disease and other forms of suffering, they resist medical intervention on the *sole* grounds of human desire. The medical necessity of saving and preserving life is distinct from the market-driven desire to manipulate life itself. As Leon Kass, the former chair of President Bush's Council on Bioethics, puts it,

> When a physician intervenes therapeutically to correct some deficiency or deviation from a patient's natural wholeness, he acts as a servant to the goal of health and as an assistant to nature's own powers of self-healing, themselves wondrous products of evolutionary selection. But when a bioengineer intervenes for nontherapeutic ends, he stands not as nature's servant but as her aspiring master, guided by nothing but his own will and serving ends of his own devising.[3]

Like Fukuyama, Kass takes aim at the utilitarian ethic of basing medical experimentation and intervention on interests (usually driven by the market) alone. The "Promethean" desire to remake human nature, simply because we will to do so, contradicts both the purpose of medicine and the natural rights of human beings.

It may be pardonable to see natural rightism as more akin to a political philosophy rather than a theory of ethics per se. According to James Ceaser, a defender of this tradition, the success and failure of natural rightism as a force tends to coincide with times of political upheaval in American history; the Civil War period is only the most dramatic example of historic debate over the validity of natural rights, although it has experienced both resistance and support in periods as different as the Cold War and the War on Terror.[4] Yet natural rightism is more than a political philosophy. It is an account of what is best and most worthy of preservation in human nature.

It is fair to say that the ethic of natural rights rests on a metaphysical account of human nature which is usually lacking in its rivals. Utilitarians presuppose nothing fundamental about human nature, beyond the assumption that our be-havior is supposed to maximize pleasure and minimize pain. For this reason, natural rightists fault utilitarians for breaking down the distinction between human beings and animals. Even Kant's ethic of duty, which also insists that we have fundamental moral obligations to treat each other as we would wish to be treated (based on a larger principle of human dignity), steers clear of making grand claims about what is intrinsically dignified about human nature. Since Kantians make ethical decisions based on the "will" to obey the categorical imperative of moral universality, natural rightists interpret this duty-based ethic as incompatible with any ethic based on an understanding of the "human es-sence." Once Kant privileged moral autonomy as the greatest good (from which he derived a duty to obey the golden rule), he set the stage for an ethic of sheer will, whose practitioners can "create" values without necessarily recognizing what is intrinsically owed to human nature.[5]

Natural rightists have no illusions that they face an uphill battle in resisting the pace of scientific change. Yet they also believe that the real battle against natural rights is driven as much by ideology as it is by the technological revo-lution. Leaving aside the tumults of the political world, natural rights theory is in "poor repute" in an academe dominated by utilitarianism, positivism, and postmodernism; all of these doctrines are suspicious of any generalizations about human nature and natural goodness.[6] Natural rightists must fight a war of ideas with their opponents so that they can convince citizens of liberal democracies to restrain the most extreme manifestations of the biogenetic revolution (especially the Promethean will to recreate human nature). As they freely admit, it has al-ways been a struggle for natural rightists to make their case even in the nation whose highest traditions are wedded to an ethic of natural rights. The biogenetic revolution, which is not strictly political or academic, threatens, in the words of Kass, to "blind us to the larger meaning of our ideals, and may narrow our sense of what it is to live, to be free, and to seek after happiness."[7]

Yet they are also confident that *reason* is on their side. In contrast, their greatest opponent, the ethic of utility, rests on premises which only appear to be rational. As Fukuyama argues, the utilitarian temptation to reduce all of hu-man behavior to a calculation of pains and pleasures profoundly ignores the sheer complexity of human nature. Indeed, the absence of a concept of what is naturally good for human beings prevents utilitarians from articulating a "hierarchy of goods," in which rational calculation can distinguish a higher pleasure from a lower one.[8]

Still, which alternative account of reason do natural rightists primarily in-voke against the always seductive appeal of utility? Natural rightists reach as far back as the classical Greek tradition of philosophy in order to make their case for a rational defense of natural right. Both Plato and Aristotle provide the necessary account of reason to combat the more vulgarized (utilitarian) version of reason in the modern age. What natural rightists often call "classical natural right" rests on the metaphysical assumption that human beings by nature have a *soul* which must be protected against the worst depredations of technological intervention. In describing the three-part soul which Plato famously describes in the *Republic*, Fukuyama applauds this account of human nature as superior to that of utilitarianism. Whereas utilitarians reduce all of human nature to the play of pains and pleasures, Plato understood that desire (*eros*) is only one part of the soul, which must be distinguished from reason (*nous*) and spiritedness (*thymos*).[9] Following in the footsteps of Plato, Aristotle also taught the irreducibility of human nature. The *telos* or soul of each human being is a reflection of more than the endless desire for pleasure. As Leon Kass puts it, "for [Aristotle] the soul was not an ethereal spirit or a ghost-in-the-machine but an immanent and embodied principle of all vital activity." This spiritual reality, as Kass argues, deserves the respect of science, and protection of its dignity against technological reinvention of human nature.[10]

It might be puzzling that natural rightists ground a modern account of ethics upon an ancient foundation. After all, the concept of natural rights in the American tradition rests on the philosophy of John Locke, rather than Plato and Aristotle. Moreover, Locke famously dismissed any metaphysics indebted to belief in the "soul." It is also far from obvious, on philosophical grounds, that an appeal to the soul of human beings can withstand the scrutiny of modern science. Yet natural rightists provide a single response to these two challenges: that the higher account of human nature, which Plato and Aristotle offered, is based on reason *and* forms the basis of the American founding. According to Fukuyama, the "dialogue" over natural right which Socrates, Plato, and Aristotle initiated continued to have an influence on the Western philosophical tradition right up to the period when "liberal democracy was born" in America.[11] John Locke, an empiricist philosopher, presumably did not escape the influence of the old classical teleological approach to nature, which his American admirers then incorporated into the founding documents of the republic.

These claims about the intellectual genealogy of natural rightism are debatable, to say the least, particularly when there is hardly any evidence that Locke or the American founders took Greek philosophy seriously. Yet natural rightists like Fukuyama and Kass insist on this "rational" foundation (at least in the Platonic and Aristotelian sense) for natural right because they must counter the accusation that their theory rests upon a *religious* basis. For this reason, natural rightists take pains to distinguish their account of ethics from another great tradition of the West, that of biblical revelation. For the remainder of this chapter, I shall examine just how successful they are in arguing that natural right is necessarily independent of a religious view of human nature.

Do Natural Rightists Need Religion?

Since natural rightists insist that all human beings possess rights by nature, regardless of culture, creed, or history, they are understandably squeamish about being reliant on a particular faith for support of their ethic. For this reason, Fukuyama rejects "divine" or "God-given" authority as the true basis of natural rights, since these rights would be unintelligible to human beings who have no historic experience of revealed religion. In particular, the history of religious wars and conflicts in modernity teach that "it is extremely difficult to achieve political consensus on issues involving religion." Rather than appealing to the authority of the Judeo-Christian God, a liberal democracy must simply look at human nature as it is; this is exactly what the first great American natural rightist did:

> Despite Jefferson's invocation of the Creator in the Declaration, he believed, like Locke and Hobbes, that rights needed to be grounded in a theory of human nature. A political principle like equality had to be based on empirical observation of what human beings were like "by nature." The practice of slavery was in principle contrary to nature and therefore unjust.[12]

It is not that natural rightists lack respect for the biblical tradition, which has undeniably influenced American politics and history well before the inception of the republic. Kass believes that both Athens and Jerusalem teach the uniqueness of human nature, as that special paradox of body and soul abiding in a cosmos not of our making. From Aristotle as well as we can learn about the spiritual and physical value of human beings. Both traditions are unapologetically anthropocentric in celebrating the distinctive nature of human rationality and virtue.[13]

Nevertheless Kass stops short of claiming that an ethic of natural right should rest on revelation *as well as* reason. Like Fukuyama, he is concerned that the morality which Scripture teaches is not always clear about the sanctity of life, a principle which underpins the very logic of natural right. Indeed, the term "sanctity of life" does not even appear in the Hebrew Bible or New Testament. Moreover, pagan and atheistic philosophy can condemn the arbitrary taking of life as vigorously as the biblical tradition. For this reason, Kass believes that the valuation of human dignity is intelligible to all human beings by nature, regardless of whether they are Jews, Christians, Muslims, or non-believers. The truth of the biblical condemnation of murder need not (and should not) be based on the Bible's authority alone. As Plato's *Euthyphro* demonstrated, the fact that a god declares an act just does not necessarily make it so.[14]

Once again, natural rightists believe that the tradition of classical Greek philosophy, as Plato and Aristotle articulated its principal teachings, is at least as viable a foundation of natural right as biblical revelation. Indeed, the fact that only the Greeks, not the prophets and apostles, refer to "nature" as a moral authority further appears to underscore their view that modern natural right is most similar to the teachings of the ancients on virtue and the sanctity of life. As a few natural rightists have argued, the American founding owes nothing fundamental to biblical principles. When the founders appealed to "nature's God," they were consciously repudiating any hint of dependence on "sacred history" or the Puritan beginnings of America. As Ceaser argues:

> Although the Declaration and *The Federalist* take note of Providence and of indications of Divine favor, the references are almost always to matters that can be confirmed by rational understanding as well. History was no longer mapped out on the basis of how events fit into a Providential plan. The Founders' concern was with the political, not religious, consequences of their actions. The aims of government were expressed in terms that address these "profane" or political ends, among them the protection of rights, security, and achieving the common good.[15]

In being the first people "to bring nature down from the realm of philosophy and introduce it into the political world as a foundation of a full nation,"[16] Americans did not need to appeal to the authority of God (although they often did). Reason did not need revelation to teach the meaning and importance of natural rights.

The assertion that the ethic of natural right does not require the persistent influence of biblical morality in the American tradition fails to stand up to scrutiny, in my judgment. In fact, the historic experience of the nation's struggles with the most tortuous moral challenges suggests just the opposite. As every natural rightist knows, President Abraham Lincoln famously wedded the cause of natural rights to the authority of God's justice in his campaign against the expansion of slavery, as well as utilitarian defenses of bondage as a matter of "interest." Even Fukuyama admits that a close cousin to natural rights, the theory of *human* rights, has Western, Christian origins.[17] To be sure, natural rightists might dismiss appeals to God as merely the rhetoric of astute leaders governing a believing population, but do Americans as a people still need to believe in a divine authority in order to buttress their faith in natural rights for all human beings?

Natural rightists like Ceaser point out that Christians in America have not always been reliable supporters of natural right, which would then suggest that appeals to revelation have often been unnecessary and even unsuccessful in encouraging faithful belief in the rights of all human beings. For example, when the rise of Social Darwinism in the late nineteenth century threatened to undermine the still fragile belief in the natural equality of all, many Protestants committed to the "Social Gospel" wedded evolutionism to the coming of the Kingdom of God and even embraced Anglo-Saxon racism in the process.[18] Moreover, since the Bible did not explicitly condemn the practice of slavery, it is far from obvious that revelation can be a useful or necessary basis for belief in natural rights. (That said, Ceaser ignores the fact that the most determined opponents of Social Darwinism were also overwhelmingly Protestant.[19])

Natural rightists are so leery of appeals to revelation that they even shy away from identifying their philosophy with the older medieval tradition of "natural law." While they are not directly opposed to all appeals to divine law, as we have seen, natural rightists see in natural law theory an unworkable attempt to synthesize the ethics of Aristotle and Scripture. This synthesis gives the false impression, in their view, that reason (in the classical pagan sense) and revelation need each other in order to function as a viable account of human nature. Natural law theory simply needs the authority of revelation too much.[20] If reason is presumably independent of revelation, then the modern heirs of Plato and Aristotle can surely use exclusively rational arguments to support the principles of natural rights.

We have seen that natural rightists prefer to invoke the authority of Greek political philosophy as *the* foundation of modern natural rights. While they acknowledge that these rights rest on a concept of nature which is indebted to modern science rather than Aristotelian teleology,[21] they nevertheless believe that their account of ethics is the best defense against utilitarianism and other modern ideologies. Still, how much does modern natural right have in common with the Platonic and Aristotelian teachings on natural right? It is a fair question, since no Greek philosopher writes of "natural rights" belonging to all human beings, although they certainly speak of natural right (the right of the strong to govern the weak). Yet the very idea of rights shared by all according to the authority of nature reveals a universal commitment to equality and dignity which perhaps is more indebted to a biblical tradition.

When natural rightists face the thorny question of the *origins* of their doctrine, they often find it difficult to look to the tradition of Athens. By their own admission, any doctrine of "the rights of man" was unknown to antiquity.[22] Indeed, the very question of human equality does not appear as an issue of debate until well into the modern era. Even in the medieval period, there was more discussion of what is universally good to all human beings, in contrast to ancient natural right, with its emphasis on what is good for human beings based on their distinctive natures. As one natural rightist explains:

> Unlike the natural law tradition, Natural Right does not culminate in a specific universal morality or a best regime valid everywhere and always. The Natural Right position is far more flexible because the focus is primarily on the existence of a hierarchy of "natural" types of human beings with the philosopher at the peak. Moral and political matters must be adjusted to ever changing circumstances in light of the need to maintain that natural hierarchy.[23]

If this description of ancient natural right is accurate, then it is unclear what connection, if any, it has with its modern counterpart. Modern natural rightists, to be sure, are not necessarily opposed to all talk of "hierarchy" in a positive sense. Kass, for example, laments the fact that the biogenetic revolution has a leveling effect on human beings in at least two ways. First, if we one day have the power to reproduce "superior" human beings in terms of intelligence and athletic prowess, the very meaning of words like "competition" and "excellence" will be compromised. The concept of "superior performance" will be artificial or even meaningless if our use of technology can level the playing field for all human beings.[24] Perhaps worst of all, the ethical valuation of inequality, the very basis of competition, might disappear:

> Beyond its everyday utility, superior performance also *ennobles* society; it makes everyone better; it raises the spirits of a community; it nourishes the desire to be better and to do better, as individuals and as a people. The example of superior performers gives those who are still developing an image of who or what they might aspire to become themselves.[25]

Second, the promise of technologically feasible ways to postpone death, ease all suffering no matter how trivial (through therapeutic drugs), and produce artificial happiness, panders to the most slavish passions within a liberal democracy: the general unwillingness to face the challenges of life and death without easy dependence on the conveniences provided by science.[26]

What Kass describes as the leveling effects of the biogenetic revolution sounds like an appreciation of hierarchy in the natural sense, and thus comes across as an endorsement of classical natural right teachings about the virtue of inequality (particularly as a spur to other virtues like self-control, ambition, and honor). Yet modern natural rightists are not unconditional defenders of all hierarchy, since they worry about the possibility of a new elite emerging from the biogenetic revolution as well. The relatively new practice of preimplantation genetic diagnosis (PGD), which was originally intended to detect genetic or chromosomal abnormalities before the start of a pregnancy, has provoked concerns about the *inegalitarian* implications of this procedure. Kass warns that PGD could create "a society divided between the economically *and* genetically rich, on the one hand, and the economically *and* genetically poor on the other."[27] The question of who is genetically "fit" may well reflect the priorities of social class. Would the ailments of the rich take precedence, based on their greater financial access?

Despite the assurance of natural rightists that their tradition would restraint the appetite of wealthy Americans for these technologies, it is not obvious that fidelity to natural rights alone can do the job here. Even Kass admits that Americans who possess the libertarian love of unlimited economic freedom, a corollary of natural rights, would resent any restriction on access to these technologies if the consumer can pay for them.[28] Moreover, the Platonic and Aristotelian defense of "natural hierarchy," even if it is based on a premodern concept of science, does not teach anything necessarily critical of the attempt to create a superior class of human beings (particularly if this class can be engineered to develop "virtuous" genes) Indeed, Plato's *Republic* rigorously supports the selection of superior individuals, who are most capable of virtue by "nature," to govern inferior beings.

It seems that natural rightists are too hasty in dispensing with the reliance of a biblical morality which is perhaps better equipped to deal with unique challenges related to the egalitarian sharing of technology. In preferring Athens over Jerusalem, natural rightists cut themselves off from one uniquely biblical teaching about human nature: that it is wrong to *idolize* human nature. The very idea of creating a new class of human beings with greater genetic endowments than others smacks of idolatry in a secular sense. To be sure, Greek myth warns of human hubris (most famously in the case of Prometheus). Still, in the American context, it is likely that most citizens of the republic will learn about the dangers of false pride from a faith tradition which has been influential since the Puritans landed.

On a more philosophical note, Athens and Jerusalem tend to differ on just how deserving of dignity *all* human beings are. No Greek philosopher ever dared say, as Saint Paul did, that there is neither master nor slave (at least in the eyes of God). While it may be an exaggeration to claim that the Bible is egalitarian, in light of Scriptural support for slavery, one can make a good case that Jerusalem is far less comfortable than Athens with the existence of hierarchies of the proud. Since there is no biblical attempt to justify hierarchy (over slaves, women, etc.) on the basis of nature, the implication follows that all authority must be justified in the eyes of God. The obligation to defend one's actions before a personal and tough-minded God of love and justice is unique to the biblical covenant; there is no god in the Greek pantheon who even remotely makes similar demands. In turn, no Greek philosopher calls upon nature or reason to question the *telos* of a slave or a master. Indeed, the natural fate of a human being may not yield to such questioning. While Greek philosophy urges moderation in all conduct, it is not particularly opposed to moderate rulers governing at the top of a natural hierarchy. Given the potentially inegalitarian effects of biogenetic technology, it is not hard to imagine a new hierarchy whose genes are more "moderate" than those whom they govern.

Thus, the classical teaching on the goodness of natural hierarchy at least needs to be countered by the other great tradition in the Western heritage. In teaching that human beings must never pretend to be "gods," Scripture also commands that God's creation must act *like* God without identifying themselves as God. This teaching is the very foundation of modern ideas of human dignity. As Kass explains:

> Yet man is, at most, only godly; he is not God or a god. To be an image is also to be *different* from that of which one is an image. Man is, at most, a *mere* likeness of God. With us, the seemingly godly powers and concerns described above occur conjoined with our animality. We are also flesh and blood—no less than the other animals. God's image is tied to blood, which is the life.[29]

Despite the fact that Kass is a prominent natural rightist, this religious language is absent in the modern tradition of natural right. Although Kass and Fukuyama insist that revelation is not the only source of teaching about human dignity, it is curious that some of the most rational arguments in favor of natural right and human dignity end up invoking Scripture (rather than Plato and Aristotle). The desire to maximize our economic and physical liberties with the purchase of biogenetic technologies is not incompatible with an ethic devoted to life, liberty, and the pursuit of happiness, although Kass sincerely hopes that natural right theory can "moderate" American desires in the marketplace; yet it is likely that the most libertarian expressions of natural rights will demand exactly what Kass opposed (as chair of the President's Council on Bioethics)—the right to practice euthanasia, prolong life through cryogenics, and create superior offspring. The force of "reason" alone may be insufficient in restraining consumerist passions, despite Kass's hopes.[30] The desire to use these technologies in order to recreate ourselves and others (including our idealized offspring) is infinitely harder to reconcile with biblical teachings. The biblical imperative to appreciate the life-blood which we have, and to be like God without using the full powers of God, is missing in the contemporary accounts of modern natural right. In short, without the biblical condemnation of idolatry, natural right philosophy may be used to justify belief in the God-like nature of human authority.

In order to get a sense of what is at stake here, I shall close this section with a brief discussion of a dramatic invocation of natural rights from the Revolutionary era. On Election Day in 1776, the famed Massachusetts preacher Samuel West delivered his sermon "On the Right to Rebel Against Governors." The foundation which inspired West to be a defender of what he calls "the just rights of mankind" was the Bible.[31] In his reading of Scripture, human beings are not commanded to obey unjust and tyrannical regimes which violate the natural right to liberty. West even went so far as to argue that the thirteenth chapter of Romans, in which St. Paul calls upon Christians to obey all regimes on the grounds that God Himself sanctions these, is opposed to obedience to tyrants. "True rulers," as West understood the spirit of Paul's epistle, are terrors to evil works. Therefore, if tyrants commit evil, they are not true rulers and deserve no obedience from human subjects.[32] What is particularly relevant here is West's confident view that reason and revelation "perfectly agree" in promoting the natural rights of humanity against the depredations of tyrants.[33] In fact, West is certain that the belief in natural rights *requires* a prior belief in God's justice superseding the arbitrary nature of human authority. For this reason, the minister counts on a people who are so opposed to the "idolatrous reverence" of arbitrary power and tyrannical government that they will jealously protect the rights which they enjoy under the God of "nature" and revelation. In particular, this people of the Book must never confuse the idols of human authority with God's justice, or the idol of "private interest" with true liberty. In calling upon his parishioners to practice the ethic of charity (love thy neighbor as thyself), West also demands that they obey only that authority which respects the rights of all human beings, without exception.[34]

It would not be controversial to claim today that belief in scientific progress persists as an "idol" of our age. The prestige of the scientific profession is often so powerful that it threatens to stifle any opposition to the long-term effects of the biogenetic revolution. As Samuel West understood, only a people schooled in the morality of the Bible can resist the idolization of any authority, political or private. While no one should be so naïve as to think that Christians are immune to the temptations of idolatry in their own lives, a culture which still takes seriously the various biblical admonitions against placing one's faith in the wisdom of powerful human authority perhaps to date offers the best chance of resisting attacks on human dignity in the name of progress or utility. Despite the natural rightists' preference for Greek political thought over the Bible as the primary source of natural rights, the vast majority of Americans have looked to Scripture rather than Plato and Aristotle for moral inspiration in their history.[35] What the most secular natural rightists fail to understand is that their own tradition requires the moral underpinning of revelation, as West taught. Reason, without the leavening force of revelation, is likely to justify as well as condemn the progress of biogenetics, particularly if it leads to new forms of "natural" hierarchy (and thus new forms of idolatry).

Admittedly, belief in the God of revelation is not the same as a philosophical argument. Nor do I offer it as a policy prescription that would solve all the challenges posed by the biogenetic revolution. Devotees of natural right are justified in worrying about the divisive effects of legislating biblical teachings as "political rights,"[36] although the tradition of natural rights has so far not been immune to similar divisions. My cautious invocation of this biblical morality should only stand as a warning that modern natural right *alone* is unlikely to prevail against utilitarian arguments in favor of prolonging or recreating life, nor is it likely to succeed against the most libertine elements of its own tradition. If our moral sense were as "naturally" inclined towards respect of rights to life and dignity as natural rightists believe, the tortuous debate over these rights would be unnecessary (just as Lincoln once noted that if equality were a "self-evident" truth, there would be no conflict over its meaning). The automatic preference for reason over revelation as the foundation of natural right may turn out to be the more dogmatic choice in this debate.

Conclusion

Nothing I have written should suggest that the choice between reason and revelation is a simple one, particularly in the complex debate over biogenetic technology. The premises of Athens and Jerusalem, which are not easily compatible, have nevertheless inspired attempts to counter the most radically modern attempts to recreate human nature. Perhaps it is hubris of natural rightists to place their faith in one tradition without giving the other its due. It is not valid to appeal to reason while pretending that revelation, although useful, is *unnecessary*. While natural rightists are confident that the political philosophy of the Greeks should remind us of the virtue of a hierarchy which celebrates excellence and competition, they should be equally emphatic that the morality of Jerusalem also tempers our appetite for false pride and God-like power. For this reason, the conflict between these two great traditions is, in the words of Leo Strauss, "the secret of the vitality of the West." It is frightening to imagine our moral compass no longer nourished by this vitality.

Notes

1. Francis Fukuyama, *Our Posthuman Future: Consequences of the Biotechnology Revolution* (New York: Farrar, Straus, and Giroux, 2002), p. 110.
2. Fukuyama, p. 217.
3. *Beyond Therapy: Biotechnology and the Pursuit of Happiness: A Report By The President's Council On Bioethics*, foreword by Leon R. Kass, M.D., chairman (New York: ReganBooks, 2003), pp. 287-8. Since Dr. Kass is the principal author of this report, I shall be using his name in reference to this work.
4. James W. Ceaser, *Nature and History in American Political Development: A Debate* (Cambridge, MA: Harvard University Press, 2006), pp. 77-79.
5. Fukuyama, pp. 119, 123-124.
6. Fukuyama, p. 217.
7. Kass, p. 309.
8. Fukuyama, p. 116. For a thorough discussion of the differences between natural rights and utility, see Joseph Hamburger, "Utilitarianism And The Constitution," in *Confronting The Constitution*, edited by Allan Bloom (Washington, DC: American Enterprise Institute, 1990), pp. 235-257.
9. Fukuyama, p. 118.
10. Leon R. Kass, M. D., *Life, Liberty and the Defense of Dignity: The Challenge for Bioethics* (San Francisco: Encounter Books, 2002), p. 294.
11. Fukuyama, p. 13.
12. Fukuyama, p. 111.
13. Kass (2002), pp. 20-21. See also Fukuyama's discussion of the defects of animal rights for reasons related to belief in the uniqueness of human dignity (146-147).
14. Kass (2002), pp. 235-241. Fukuyama also believes in a "natural moral sense," common to all human beings (142).
15. Ceaser, p. 17.
16. Ceaser, p. 22. For a similar argument which downplays the Protestant origins of the Founding, see Michael P. Zuckert, "Natural Rights and Protestant Politics," in *Protestantism and the American Founding*, edited by Thomas S. Engeman and Michael P. Zuckert (Notre Dame, IN: University of Notre Dame Press, 2004), pp. 21-75.
17. Fukuyama, p. 113.
18. Ceaser, p. 58.
19. See Richard Hofstadter, *Social Darwinism in American Thought* (Boston: Beacon Press, 1955), p. 86.
20. See Larry Arnhart, "Roger Masters: Natural Right and Biology," *Leo Strauss, the Straussians, and the American Regime*, edited by Kenneth L. Deutsch and John A. Murley (New York: Rowman & Littlefield, 1999), 296. There are a few natural law theorists who try to avoid appeals to faith, with mixed success. See John Finnis, *Natural Law And Natural Rights* (Oxford: Clarendon Press, 1980).

21. See Leo Strauss, *Natural Right and History* (Chicago: The University of Chicago Press, 1953), pp. 7-8.
22. Harry V. Jaffa, *A New Birth Of Freedom: Abraham Lincoln And The Coming Of The Civil War* (New York: Rowman & Littlefield, 2000), p. 375. Oddly, Jaffa portrays Aristotle as a critic of slavery, despite his admission that an equivalent of modern natural rights did not exist in antiquity. In the debate over abortion, natural rightists have occasionally invoked Aristotle as a critic of abortion, an unlikely position for a Greek whose civilization tolerated infanticide. See Hadley Arkes, *Natural Rights & the Right to Choose* (New York: Cambridge University Press, 2002), pp. 5, 68. See also the review of Arkes by James R. Stoner in "The Genteel Abolitionist," *Claremont Review of Books* Spring 2003.
23. See Gregory Bruce Smith, "Athens and Washington: Leo Strauss and the American Regime," in Deutsch and Murley, *Leo Strauss*, p. 119.
24. Kass (2003), pp. 103, 133, 144-5.
25. Kass (2003), p. 152 (his italics).
26. Kass (2002), pp. 45-46, 50-53; see also Kass (2003), pp. 284-5.
27. Kass (2003), p. 51, and pp. 281-83.
28. Kass (2003), p. 282.
29. Kass (2002), p. 242 (his italics). See also Kass, "Human Frailty and Human Dignity," *The New Atlantis* Fall 2004/Winter 2005.
30. Kass (2003), p. 309. See also Kass (2002), p. 233. Significantly, in his work on the Hebrew Bible, Kass is reluctant to rely on "unassisted reason" alone to explain the foundations of morality. See his *The Beginning of Wisdom: Reading Genesis* (New York: The Free Press, 2003).
31. Samuel West, "On the Right to Rebel Against Governors (Election Day Sermon," in *American Political Writing during the Founding Era 1760-1805,* Vol. I, edited by Charles S. Hyneman and Donald S. Lutz (Indianapolis: Liberty Press, 1983), p. 427.
32. West, pp. 426-427.
33. West, p. 431.
34. West, pp. 415, 431, 442-444.
35. See Clinton Rossiter, *Seedtime of the Republic: The Origin of the American Tradition of Political Liberty* (New York: Harcourt, Brace, and World, 1953), p. 356. Rossiter notes that Plato "was virtually ignored" during the Revolutionary period, while Americans relied heavily on the Bible for guidance.
36. Fukuyama, p. 111.

Taking Life: Science-Based Justifications in the Third Reich

Christopher Vasillopulos

Prologue

Many years ago a short, very short Hispanic woman, perhaps in her thirties came into my office. She was a new advisee and a town select-man. Although we differed on nearly every political and social issue, we became friends. She took and excelled in every course I taught, often to the dismay of feminist colleagues and students. After what she had been through, this sort of disapproval was nothing.

She was born with a horrible bone defect, which required scores of operations to adjust the growth of her bones to the rest of her body. The love and devotion of her family undoubtedly was essential, but, at least in my view, what got her through these multiple ordeals was her courage and her determination to be someone.

In a quiet moment, I said to her, in the manner of a confession, "I would have aborted you."

"Yes, I know. And if I had known all the suffering I was to endure and all the sacrifice I visited on my family, I would have aborted myself."

"I was wrong."

"I am glad you think so."

Yolanda has lived a life of service to her community, holding virtually every post and office her town has to offer. She has won numerous state and national honors. What is most important to me, however, is that her life has changed my ideas about life and its qualities.

Taking Life: Some Preliminary Existential Problems

"Those who take lives, who see in their ideas a rationalization for mayhem and carnage, are no less taken with ideas than are the life givers. Hence, a culture of life taking no less than life giving is wrapped up with the essential social institutions and cultural agencies of our age."— Irving Louis Horowitz[1]

"In extreme and intense fashion [Hitler's dictatorship] reflected among other things, the total claim of the modern state, unforeseen levels of state repression and violence, previously unparalleled manipulation of the media to control and mobilize the masses, unprecedented cynicism in international relations, the acute dangers of ultra-nationalism, and the immensely destructive power of ideologies of racial superiority and ultimate consequences of racism, alongside the perverted usage of modern technology and 'social engineering.'"—Ian Kershaw[2]

Without committing to the entirety of his philosophy, much of this section derives from my reading of Spinoza. First, all human actions operate under a cloud of ignorance. Although its degree varies widely and significantly, the ethical implication of "partial knowledge" is that no one can know in principle whether a given act is good or bad in the final analysis, *sub specie aeternitatis*. Only God or Nature "has" all the facts, past, present and future. The scare quotes are meant to alert the reader to anthropomorphisms which were anathema to Spinoza. God does not possess anything. In a sense God *is* all the facts. Given this understanding, Spinoza's metaphysical determinism is difficult to refute.[3] Yet, due to partial knowledge humans are "free" to act (or seem free), no matter how much their actions have been "caused," including a pre-existent "causal" chain. Since causality operates within time and space, I am not sure how to deal with it as "pre-existent." However this may be, human beings are existentially "responsible" for their actions, not withstanding the absence of Aristotelian choice. At the very least, they are held to be responsible for their actions by their societies. Socially defined criminal acts can therefore be properly punished without embarrassment, despite metaphysical determinism. The logic is that of killing a rabid dog. No one believes the dog is *guilty* of anything. Considered a danger to society, he can be killed in the name of prudence. Spinoza means more than "life is not fair." He means that concepts like "fairness" or "criminal" are social labels derived from prudential concerns.

Allow an example. No one could have known (or know now) in principle whether assassinating Hitler, say in 1938, would have been a "good" thing to do. I do not here mean, because no one could be certain the murder of millions of Jews and others would occur. I mean that even if all this were known, no one could be sure that killing Hitler would have been "good" or even better than any alternative. I believe that prudence would have indicated that Hitler and his Movement contained the seeds of genocide and that his elimination would have been desirable. This conviction, however, can make no compelling claim to knowledge or goodness. It can claim prudence, but no more. Spinoza did not apply Aristotle's notion that knowledge can be a proposition that is true "for the most part." At the same time we can and must act in "partial knowledge," regardless of our ignorance of the facts or the moral accounting based on them.

Nevertheless, the absence of the possibility of knowledge in principle should at the very least give us pause when we contemplate any irrevocable act like the taking of life. If we cannot be sure of the taking of Hitler's life—the current paradigmatic case—then less certainty must apply to other cases. It should be noted that "paradigmatic" is merely a social label and carries no metaphysical significance. Yet we (our societies) take life, if not routinely, frequently and often without a great deal or any reflection, anguished or otherwise.

1. In war strangers kill each other as a matter of duty if not glory. The justification of taking life in war is political, not individual. Even in a purely defensive war, the taking of life cannot be entirely attributed to the enemy. All military commanders and their political superiors engage in military triage. They make almost continuous assessments regarding how many soldiers will die to meet a given objective. In other words, the question is, How many soldiers can be "properly assigned" to death at least probabilistically? In the best circumstances in the sense of the most justifiable, the few are sacrificed for the many or for a larger "good." William James makes this point with characteristic flair: "We, the lineal representative of the successful enactors of one scene of slaughter after another, must, whatever more pacific virtues we may also possess, still carry about with us, ready at any moment to burst into flame, the smoldering and sinister traits of character by means of which they lived through so many massacres, harming others, but themselves unharmed."[4]

2. In virtually all societies, police are entitled to take life to secure the lives and property of the public, as well as, to defend themselves. When the police cannot be sure of the danger, they can be excused in some circumstances from punishment for taking life, if they can show they acted in good faith and in a reasonable manner. It should be added that self-defense allows for the taking of life in general.

3. Many societies employ capital punishment as a criminal sanction. These societies take life in cold blood, that is, without the existential problematics that normally surround police actions.

4. Although the status of a fetus remains controversial, at some point all agree it becomes alive, that is, viable outside the womb. As medicine advances, this point approaches conception. Hence legal abortions at some point have the same moral, if not legal, status as late-term abortions. Thus whatever the justifications—the health of the mother, the prospects for the child, the constitutional principles of privacy or equality—abortion can be seen as the taking of life. It must be remembered that there can be no scientific basis for weighing the life or health of the mother against the fetus. These evaluations are social and political, whatever the moral or ethical arguments which claim to underlie public policy. How powerful the social and political element is in matters of abortion can be seen by the nearly universal acceptance of the taking of the life in cases of incest or rape. No scientific basis need exist at all to terminate a

pregnancy under these social conditions. In other words, should the scientific evidence be conclusive that the fetus is a human life from conception, it would not be determinative of the issue of abortion. No feminist group would outlaw the taking of life on this ground, nor would most societies outlaw it in instances of rape or incest.

5. Furthermore, it should be understood that such decisions regarding the relative value of life are routinely determined in every hospital budget. Should we devote more resources to emergency care, when it cuts into research allocations? Should we have more kidney machines or more cat scans? Every public budget involves the same matters. Should we spend more money on hospitals or schools, on police or teachers? So does every public policy. Should we increase the retirement age to pay for higher education? Should we ration medical resources for people over the age of 80, 90, or 110? Should we allow the rich unlimited medical services or unlimited use of other scarce commodities, merely because they have the money to pay for it? And so on.

6. Let me here deal with some issues of life-taking which may seem trivial. About 30,000 Americans die each year in traffic accidents. Nearly all of them are preventable. Reducing speed limits, making safer cars and roads and many other policies would undeniably and drastically reduce traffic deaths. To the degree that the preventable is not prevented, they are "allowed." They are "allowed" because society had deemed other matters more important than human life, though no one quite puts it this starkly. Moreover, the only serious discussion for limiting the lethality of motor vehicles has centered on the conservation of oil or the protection of the environment. Finally, a word should be said about other preventable deaths due to unhealthy life styles, like smoking, alcoholism, obesity, stressful jobs among others. Clearly lives are "taken" (or "given") in the name of dubious pleasures or addictions or seeming necessities.

In conclusion, while there is much talk about the preciousness or the sanctity of human life, no society suggests that life is an absolute value transcending all others. Life is one value among others, some which seem significant to many of us, for example, freedom or honor, some of which seem matters of convenience or desire, like driving fast or eating donuts. From this perspective I believe a reasonable discussion of the scientifically grounded taking of life can proceed. First, however, we have to deal with the problematics of science and society.

The Problematics of Science and Society

"National Socialism is a cool and highly reasoned approach to reality based on the greatest scientific knowledge and its spiritual expression…"—Adolf Hitler[5]

"In line with many advocates of eugenic sterilization and euthanasia…, Hitler believed that anyone not fit for life should perish, and that the state should give a helping hand…. [As Hitler said]: 'the bloodiest civil wars have often given rise to a steeled and healthy people, while artificially cultivated states of peace have more than once produced a rottenness that stank to high Heaven.'"—Michael Burleigh[6]

First, science cannot *determine* policy no matter how grounded a given policy is in science. The significance of science to the policy at issue is not a scientific question but a poetical one. Nor can the worth of the policy be measured by science. Science cannot say that a hospital is more important than a football stadium. Second, science is in itself problematic—that is empirical and probabilistic. No scientific finding or premise can be forever or absolutely true. This essay, however, will assume that well-established science is true or true enough. In other words, the probabilistic and empirical basis of science will *not* be used to limit its application to policy. Moreover, this essay will use "science" to refer to statements about reality which do not as yet have a scientific basis. For example, I will assume the genetic code has been cracked and that we can "know" how an individual will turn out within the parameters that public policies normally assume. One purpose is to make the case for the scientifically grounded taking of human life as strong as possible. Another is to move policies for taking life away from the convenience or desire end of the spectrum and toward the self-defense or preservation of society side.

Let me suggest some examples which as yet do not have a scientific basis, examples whose likelihood far exceeds the probabilities normally considered appropriate for preventative social action: we can genetically determine, for example: (1) who will be a serial killer; (2) whether a fetus that will live a life of unendurable pain; (3) whether a child will be severely mentally impaired; (4) whether a child will be unproductive and be a drain on the resources of the society to the point of impairing the chances of survival of "normal" children. In other words, the expenses of caring for such a child will cut into hospital budgets to the point that other children will be put at risk; (5) whether a fetus will develop a disease that will kill thousands before a cure is found.

To sum up, in these cases (and others like them) the question is, Can society be justified in taking life (or withholding the services necessary to life) for a "defective" child or fetus to further the public good? The reflexive answer is, "No." The emotional response is to avoid the issues such questions thrust upon us. Yet in the context of taking life as already discussed, can such an irrational response be supported? Or stand unexamined? If so, then how can one deal with the vexed topic of abortion, which at some point in fetal development becomes a life, a life taken in the name of some more important value? Even if one takes the position of the absolute right to life, what about instances when the mother's life is at risk? If one softens the absolute position by allowing definitions of *human* life to qualify organic existence, more difficult issues arise. I do not wish to examine the controversies which swirl around the issue of abortion. I raised the issue in an effort to avoid the cant and sentiment which often plagues discussions of taking human life.

Again, let us assume that scientifically demonstrable incurables and unproductive human beings exist. "Demonstrable" here means "knowable within the parameters of probability the society has determined is acceptable for the formation of scientifically grounded public policy." In other words, let us assume that science is not problematic, that there is no problem of applications. Let me also assume that the "slippery slope" argument does not apply *within the realm of science*. This means that scientists acting in their capacity as scientists will not soften and will not need to soften scientific canons to serve other values. For purposes of argument, I am assuming science and scientists are more pure than they are.

The "slippery slope" difficulty remains, because non-scientists will be tempted to move beyond the scientifically demonstrable cases to those more doubtful to meet political or social needs. Society, notwithstanding these optimal scientific circumstances, remains problematic, because society is itself a socially constructed term. Defining itself and its prerogatives leads to all sorts of concerns, many of which have vexed political theorists and jurists for millennia. Unless one vests "society" with organic properties as a kind of life form, it is merely an abstraction which covers the decisions of a small group which impact a large group of people not all of whom are within its jurisdiction. Many social scientists, who could not be classified as Nazis, come close to this "organic" view. Consider Leslie White: "Instead of regarding the individual as a First Cause, as a prime mover, as the initiator and determinant of the cultural process, we now see him as a component part, a tiny and relatively insignificant part at that, of a vast socio-cultural system that embraces innumerable individuals at any one time and extends back into the remote past as well...."[7] This "vast system" has its own logic and justification. The naturalness or the inevitability of human clumping, call it what you will, does not invest the *group* or the *system* with organic, mechanistic, or logical qualities, independent of its members.

There is another problem with the socio-political context of science. Powerful or rich societies impact less powerful and less wealthy societies by the very nature of their activities. When it is appreciated that scientific knowledge and its technologies are virtually isomorphic with wealth and power, the problematic of society in a world of nation-states becomes obvious. For example, suppose a given nation decides to treat its incurables and its unproductive members humanely, but can only do so by exploiting its neighbors. This may seem shocking, but it has been the basis of many imperialistic policies and aggressive wars throughout human history. Moreover, such practices have been justified by nearly all "Realist Theories" of international relations. A nation-state is responsible for the well-being of its citizens, not for anyone else. It cannot jeopardize the well-being of its citizens or its self-preservation in the name of any other value, without violating its reason for being. Whatever one thinks of the justification of self-preservation of individuals or states, there is little doubt that the principle has governed international relations for thousands of years. In war, many political and military leaders have said that the life of one of their soldiers is more important than the entire enemy, which includes all their women and children. Virtually all of them have acted under this horrific truth.

This position may seem outrageously stark and oversimplified. Yet ask yourself what sacrifices would you make to improve the lot of the citizens of other nations, to say nothing of nations hostile to America? The global warming problem and other environmental issues have foundered on this shoal of self-interest or national self-interest for decades. Nation-states have been willing to spend billions on defense and war and virtually nothing on dealing with their underlying causes in other countries. Twelve billion dollars a month in Iraq has been spent for years, without any serious discussion of spending one ten-thousandth of that per year on economic development in poor countries. Only recently, after many years of futility, have the opportunity costs of the war been discussed in America for Americans. I don't believe Americans are worse than other people. I make these rather prosaic points not to cast aspersions, but to create a context (not an equivalency) for my discussion of Nazi justifications for the scientifically-grounded taking of life.

Nazi Solutions: Taking Genetics Seriously

"No consistent eugenicist can be a laissez-faire individualist unless he throws up the game in despair. He must interfere, interfere, interfere!"—Sidney Webb[8]

"The judge must always bear in mind Hitler's words that 'the right to personal freedom always gives way to the duty of preserving the race.'"—Hereditary Health Court Judge

Every treatment of the Third Reich devotes at least some space to euthanasia and eugenics, if only to prepare for a discussion of the mass murder of Jews and other "undesirables." Of course, from the Nazi point of view, killing Jews was a form of eugenics or in their terminology, "racial hygiene." It is significant to note that the Nazi made concerted efforts to justify their racial policies scientifically. In keeping with my method for discussing the scientific basis for taking life in its strongest terms, I will ignore for the moment "racist science" and the Judeocide it helped to legitimate. The racial basis of National Socialism did not (and in my view cannot) have a scientific basis strong enough to justify a rational public policy. Nevertheless, the Nazis did believe their anti-Semitism had a biological basis. It should also be understood that the biological basis of society, and thus the science that understood it, was the basis for all Nazi social values. Nature (and science) was not only implacably there. It was implacably correct. Moreover, they knew that the aura of science would help to dull the resistance to their treatment of the Jews. It is useful to discuss euthanasia first for three reasons: these programs antedated the Judeocide, were prior to it in logic and were a politically and socially necessary preparation for it.

The Nazis are often assumed to be ham-handed thugs, sociopaths, who killed for the joy of it. The problem with this assumption with respect to euthanasia is that it cannot account for the debate which surrounded their programs for the elimination of those "unfitted for life," debates which took place in many nation-states, including Britain and the U.S. Nor can it account for the implementation of such programs by well-educated health professionals. Tens of thousands of people were killed in the name of scientifically-grounded public policy.[9] To what extent were Nazi policies for taking life rational? Can their rationality provide justification? Anticipating my argument, if one takes the omnicompetent, all-powerful nation-state as sovereign, as a self-justifying political entity, which must pursue its own interests in a hostile world, then I believe there is no way to contest Nazi policies of euthanasia, no matter how repugnant they seem. The omnipotent and omnicompetent state-based argument does not require scientifically based policies or even rationality. Its only test is success, that is, the preservation of the nation-state. Science and its effect of conferring rationality on public policies become necessary only to the extent legitimacy is required to have the policies carried out. Science here operates as a cover of rationality for policies which might otherwise create hesitation or opposition.

Science and its logic may be determinative for scientists and those elites who believe in them and for the societies who base policies on them. The masses are suspicious of what they do not understand. They fear that some scientific project may consign them to the category of the "incurable," or "unproductive." They need to believe policies make sense in less rigorous terms. At the same time, they want to believe that unpleasant decisions are out of their hands and that their "betters" will take the responsibility of these necessary programs. Thus, the scientific argument, insofar as it was aimed at any group in German society, targeted intellectual, academics and jurists and above all the SS, who were generally charged with policy implementation. I do not mean to suggest that science was used *only* as a Machiavellian instrument. Many educated people believed in science as an intellectual endeavor *and* in its application to social and political problems. Although these predilections were particularly strong in Germany, many other European and American elites had the same views, most of which were effects of the explosion of scientific knowledge, including genetics, in the nineteenth century. Of course, it takes a leap to move from a generalized respect for science and its applications to programs which take life in its name. How and why the Third Reich made this leap can now be discussed.

The Nazis were aware that euthanasia was a problematic term. Part of their concern was based in the still-Christian German society; part in the normal worry of ordinary citizens for their own lives and their inability to "guard the guardians." So the Nazis began with what they believed were the easy cases, which they well publicized. "Easy cases" are here defined as those which the public would see as fitting well within the normal categories of taking life as sketched above. I do not mean to suggest that I agree that the taking of life is ever an easy decision.

The first category of easy cases takes the concept of euthanasia literally, that is, a "good death" or a death in the interests of the person whose life is taken.[10] In a sense, the individual decides that his or her life is no longer desirable, that death is preferable. A person may rationally prefer death to a life of unendurable pain. A person may prefer death to a comatose existence, a preference perhaps indicated in a living will or conveyed to someone with the person's interests in mind. Many religions find such decisions problematic or sacrilegious, few can contest their rationality, especially if backed up by scientific findings regarding curability or the quality of the life expected. Rational problems arise when the determination of a life not worth living is made by another, a person who may not have the interests of the "patient" as a central or determinative factor. This person may be a parent who simply cannot or will not cope with a child who is severely impaired. To this "person" may be an institution that does not wish to spend or cannot spend scarce resources on such a patient. When these choices are supported by science, a great deal of the anguish may be removed. Again, this sounds more harsh (or exceptional) than it is. For example, let us suppose that an octogenarian (or pick an age that matters to you) requires a million dollar operation to live for another year or two, the same million dollars that would keep, say, 100 dialysis patients alive for the same period of time. In a society devoted to private property, it might matter whether the octogenarian could pay the million dollars. Yet even in this case, is the decision easy, given finite medical resources? Should medical resources be skewed toward the desires of the affluent?

Less easy cases deal with lives not worth living when the determination is entirely the society's, or at least, independent of the intended victim. Consider, for example, the severely mentally impaired, beginning with those who cannot take the simplest care of themselves. The Nazis made much of such cases, speaking about those who live in a sandbox, urinating and defecating on themselves. Should society be expected to provide round the clock care for such people, while it allows other social needs, like schools or nutritional services, to go unmet? Remember how viability of a fetus is often determined: can it live on its own? What about those out of the womb who cannot live on their own in the most fundamental senses? What about those who have "incurable" penchants for sex with small children? What about those who brutalize children or women due to a scientifically determined defect? What about alcoholics or drug addicts, whose habits are so severe that they drive them into criminal activity? Should society tolerate or support them indefinitely? Should it do so regardless of its opportunity costs? Should it provide the same services when it is fighting a war or undergoing an economic crisis? These questions and many others like them were asked in the Third Reich and the answers were often forthcoming in the negative. I suspect that their answers would be echoed by many Americans. As we have discussed, the role of science in these matters is not and cannot be in principle determinative. Nevertheless, the status of scientific knowledge in modern society is so high that scientific support for assessments of incurability, desirability or cost will inevitably play an important role in any public policy decision.

Now, let us deal with a much more problematic set of issues. What about the elimination of undesirable groups? On the one hand, the probabilistic nature of science would seem to apply to groups with more validity than to individuals. It might be accurate to say the average height of a group of humans is five feet eight, without any individual being precisely that tall. If it were a crime to be five eight, it would therefore be wrong to condemn any individual in the group, even if it were scientifically determined that humans five feet eight are a danger to society. One might very well think that a group that averages a given height might have more individuals who have that height than a group that averages four feet, but one could not justify punishing the entire group for this propensity. It is simply a statistical fallacy, to say nothing of morally reprehensible, to apply a group characteristic to an individual who may not have the characteristic. This is the very definition of prejudice.

On the other hand, let us assume that we know that a given group of people is statistically likely to carry a deadly virus, but that the virus is undetectable until it is too late to combat it. In other words, once the virus is detectable it is impervious to treatment. Detectability implies death. What then can be done about the group, fully understanding that many individuals in it will not have or develop the virus? I realize this may sound like science fiction, yet its logic is applied routinely by states, particularly when dealing with terrorists or other deadly enemies. "We know terrorists inhabit this village. We cannot ascertain or apprehend the individuals, yet an air strike will kill all the terrorists and save countless lives of our citizens. What do you expect us to do, hold a trial?" There is not a state in the world that would not apply this logic. That the Third Reich did so routinely and honestly would not dissuade any leader presented with a similar problem, no matter how carefully he or she might distinguish his predicament from those presented to the Nazis.

Did the Nazis believe that the Jews presented this sort of danger to them, a danger scientifically grounded? The short ideologically based answer is, "Yes." A more complete answer is, "It depends." The simpler and more easily refuted answer is that when the Nazis claimed that the Jews were "vermin" or a "bacillus," and meant the charges literally, they were making an empirical claim, one that could be confirmed or refuted by science. Such claims were not confirmed or seriously investigated. If the terms of opprobrium were taken less literally and more metaphorically, then the matter became more complex. Suppose the Jews, taken as a "race," had a series of empirically verifiable characteristics, perhaps as objective as height, and suppose these characteristics were perceived by the society as a danger to its existence, would the society be justified in taking extreme measures to eliminate the threat to its existence?

The Nazis believed that the Jews were a biologically distinct race, disagreeing only over their human status. They also believed that this race was biologically *anti-Aryan*. We cannot go into Nazi metaphysics here. Taking the Nazis at their word, that is, assuming the sincerity of their beliefs, what can we say about their treatment of the Jews which culminated in the murder of almost six million? Biologically, the concept of race has very little scientific grounding. Sociologically, the basis is more valid, that is, there are empirical generalizations that are more or less accurate that apply to the Jews in Europe as a group which distinguish them from non-Jews. In terms of this discussion, the Jews were seen by Nazis as a group of people who produced undesirables, like alcoholics or drug addicts (or dealers) and especially sexual predators, who therefore threatened the integrity and therefore the survival of Germans. Few Nazis believed all Jews were so disposed, but like terrorists in a village or a group that carries an undetectable virus, most Nazis believed that it was wise to eliminate this danger from German society. Although it cannot be determined how many of them would have sanctioned the Judeocide, it can be safely said that few would have objected it, if they perceived the Jews as a danger to their existence.

More problematic is the case when the group is defined metaphorically. Not only is the group's virus a metaphor, but its members are metaphors. "Jew" in the Third Reich meant far more than Jewish human beings or, as some Nazis would prefer, organisms. "Jew" conveyed a set of concepts which the Nazis believed jeopardized the German *Volk*: materialism, capitalism, greed, anti-Christianity, bolshevism, anti-spirituality, sexual promiscuity, physical defilement, organic weakness, congenital ugliness, and so on. This metaphor had to be annihilated. If organic Jews were in the building which had to be burned to destroy the threat, so be it. "Do you expect us to hold a trial?" Whatever the strengths and capacities of science, to employ it to deal with metaphors cannot be scientific. Irreducibly, to deal with the Jews as a metaphor called for a moral determination. Convinced National Socialists resisted this clear-cut requirement.

Instead they, including many scientists, blurred the lines between science and social science, between questions which can be subject to positivistic verification and those matters which are not but which may be true for the most part. The Nazis took advantage of the preferred status of scientific knowledge to deal with matters more properly in the province of social scientific generalization. In other words, they allowed the status of the Law of Gravity to flow into the

Law of Race. And the Law of Race was assumed to be scientifically (and, as we shall see, eternally) valid and, therefore, determinative of all rational policies dealing with the Jews. In other words, questions of morality were subsumed by what Hitler called the Life Philosophy:

> A morality based on the demands of life is unable to set up an unchangeable moral code, because the central flux of life necessitates a progressive internal readjustment. The ethics of the Life-philosophy cannot and will not provide anything but an orientation, an attitude towards these problems. It is of little avail to educate a man according to rigid, preconceived rules; the one important thing is to open his mind and to penetrate every fiber of his being with the current of life. Increase of vitality, that is, the supreme demand of the Life-philosophy.[11]

More precisely, they allowed the appreciation of the germ theory of disease to reinforce the social antagonism between the dominant population and a despised minority. This is not to say that all the Nazis were insincere or used science as a mere Machiavellian justification. Many Nazis, including scientists, took the scientific status of race seriously. At the same time, it seems clear that without a general respect for scientific authority, the Judeocide might have created more resistance, at least among elites. I believe this point is more powerful when placed in the context of the problematics of both science and society, as understood my many German intellectuals. The concept of objectivity, as the West tended to conceive it, had been considered superficial by many German philosophers for more than a hundred years, in much the same way that the West had been considered overly materialistic by many German writers and thinkers. Consider Martin Heidegger:

> If only our totally superficial culture of today, which loves rapid change, could visualize the future by turning to look more closely at the past! This rage for innovations which collapses foundations, this foolish negligence of the deep spiritual content in life and art, this modern concept of life as a rapid sequence of instant pleasures…so many signs of decadence, a sad denial of health and of the transcendental character of life.[12]

Science, the most powerful expression of the concept of objectivity, was likewise suspect by many German scientists, including Noble prize winners. To realize its full potential, science had to be informed by values, for some this implied a sense of eternal truth, for others, a social context. With its positivist discipline removed, science could then serve in good faith and conscience the eternal needs of the Aryan race and the current needs of the German people. Hitler made this point explicitly: "We will rebuild our *Volk* not according to theories hatched by some alien brain, but according to eternal laws valid for all time."[13]

By blurring science and social science, by confounding science with morality, educated Nazis, including scientists, avoided the moral question. "Do the Jews, conceived organically or metaphorically, present so great a danger to the survival of the German *Volk* that killing them or otherwise eliminating them is justified?" Much as I would like to provide a categorical answer to this question, answers must be premised on a particular value-set. My personal views, based on the concept of natural rights, are virtually categorical: a threat to life is *not* sufficient *for me* to justify mass executions, whether or not they take place under the color of law. Unless a political organization adopts natural rights premises, however, they cannot be assumed to be the basis for policy. With regard to the taking of life, the continuum runs from "under no circumstance can the taking of life be justified" to "the taking of life for amusement or convenience." Most modern societies seem to be well in the middle of this continuum, as our discussion has suggested. Lives are taken in a multitude of ways which may be justifiable only when another value or group of values is at play. So the issue turns on, what other values?

Let us consider taking life *within* a society or political community. Normally a society, as opposed to an individual acting in an existential situation, cannot take the life of someone in its jurisdiction without application of law or due process. How powerful this restriction is in fact depends on many conditions. Lawyers can always dispute any term, e.g., jurisdiction. The principle, however, is clear and generally undisputed. Usually, the law applies only to acts, not beliefs or predispositions. Science in these instances has little role beyond forensics. For simplicity I am leaving aside all those social practices which allow life to be perilous: dangerous traffic, unhealthy food, etc.

What about the taking of life between societies or nation-states? Justifications for war far exceed the taking of the life of a convicted criminal. Every man, woman, and child is the enemy and at risk. Again, science has little role here beyond the technology of applied violence. Consider Bomber Harris remarking on the militarily gratuitous fire bombing of Dresden: "I do not personally regard the whole of the remaining cities of Germany as worth the bones of one British grenadier."[14]

What about a "war" *within* society? I do not refer to civil war as it is generally understood, but the war against a presumed subversive or anti-social group? Many societies have had such internal "enemies;" some still do. America has had various Red Scares, in which radicals or Communists have been persecuted and prosecuted for beliefs. America is currently going through a terrorist-Muslim paranoia, which similarly has meant the suspension of many constitutionally guaranteed liberties. The case is always made, in one form or another, with varying degrees of sophistry, that the Constitution is a "luxury" or a "relic" which cannot properly stand in the way of survival. The question—survival as what?—is invariably begged, relying on the fear of physical death. Arguably, the Constitution protects anyone within its jurisdiction, as people who have been constituted by the Constitution, American citizens, or people who have been brought under the umbrella of its protections under its provisions. In other words, survival *as creatures of the Constitution* is what it means to survive as an American or as someone in American jurisdiction or protection. Under this interpretation, by no means universally ascribed to but with a powerful heritage in judicial decisions, there can be no war against internal enemies. There can only be prosecution for criminal activities. There is no suggestion here that violations of constitutional rights are on the same plane as taking organic life. Nevertheless, many societies have urged their citizens to take life and to risk their own lives when they were not physically at risk. The justification usually turns on some notion of the quality of life phrased in nationalistic terms, e.g., to protect the American way of life or freedom. My preference for absorbing risks based on the Constitution and its expression of natural rights is on the same logical plane as patriotic values.

How can the Jewish experience in the Third Reich be understood from this perspective? Consider a prosaic example: the Nazi response to department stores. If Jews did not invent large-scale retail business, they might as well have. In Germany Jews owned almost all department stores. Like any capitalist innovation, department stores implied the destruction of less efficient modes of distribution. Many small shops and stores were placed in economic peril, if they survived at all, German (ethnic) owned stores. Many Germans who patronized department stores were sympathetic to the plight of neighborhood stores; few, however, were willing to pay higher prices to subsidize them. The Nazis, ever sensitive to populist resentment, demonized department stores as an especially pernicious Jewish invention which would destroy the German way of life. While there is no way, I presume, to determine scientifically the evil of department stores, it is possible to come to the determination that they harm the society. It is perfectly legitimate to value one group of businessmen over another on socio-political grounds. It happens all the time in ostensibly free market, capitalist economies. This valuation can be supported by social scientific analysis which might favor one policy response over another. The Nazis could have outlawed department stores. They did not. Alternatively, they could have outlawed Jews. They did not. They killed them but *not* under the cover of law. Finally, they could have outlawed "Jewish" department stores. They did not. The scare quotes do not indicate the difficulty of defining who is a Jew? They signify the difficulty of defining a *"Jewish department store."* Is it a store which has good prices? Exceptional merchandise? Serves only Jews? Exploits only Gentiles? Makes exorbitant profits? Especially at Christmas? Is it a store where Jews can express their commercial talents? Their greed? Is it a store which incarnates the Jew as metaphor?

The Nazis dealt with Jewish department stores in a characteristic fashion. They allowed the stores to perform a needed and valued economic function, while they decried their Jewishness. Jewishness sometimes meant ownership, so they applied pressure on Jews to take Gentile partners or to sell out to them.

Jewishness sometimes meant the effect of the stores on less efficient businesses. Very little beyond rhetoric was done to ease the difficulties of small store owners. Jewishness sometimes meant efficiency. This may seem less odd when efficiency is seen as a metaphor for materialism or impersonal calculation. Efficiency became pernicious when it was associated with anti-spiritualism or an attack on the spirituality of the German *Volk*. Again, very little beyond rhetoric could be done to attack one of the signs of modernity, especially when efficiency was at a premium in a society at war or preparing for war. Most importantly, department stores were conceived by the Nazis as a metaphor for Jewish materialism, proof that they could not coexist with Germans, much less be members of the *Volksgemeinschaft*. In addition, the Jewish department store proved that Jews did not care about Germans, proved that Jewish profits were more important than decent Germans. Finally, the very success of the department store indicated a way to deal with the Jewish Question. The concept of the department store could be *purified* of its "Jewishness" and thus serve proper Germans properly. Jewishness, though insidious and persistent, could be purged. This purification process would be difficult and take a long time, but it could be done. Jews as metaphor, of course, but if Jews as organic beings were caught up in the purge, so be it.

Jewishness could not or, at least, was not criminalized, so it had to be fought as a war against an evil group or an evil set of behaviors and influences: Jews as enemy combatants. As such they were outside the law outside the community of care and, above all, outside the *Volksgemeinschaft*. Not only could they be dealt with as all power states tend to deal with obstacles to their objectives, they could be eliminated *as an ideal*. While this ideal was never broadly accepted by the Germans as a whole, it was countenanced by a sufficient number of them to allow for the murder of millions of Jews. The Jew as an organic being became one with the Jew as an anti-German metaphor, a bacillus to be annihilated for the sake of the *Volk*.

Science was a critical part of this process. When science buttressed other reasons for euthanasia, by linking "incurables" or "undesirables" to heredity, many Germans believed taking life was warranted. The science trumped traditional, pre-scientific, values. Social science, especially economics, buttressed these policies. The exigencies of war reinforced them. With young, healthy men and women dying in the millions, how could society be expected to expend resources on the severely and incurably damaged? Working together science and social science served to dull traditional Christian values regarding human life; life and death were no longer God's dispensation, but Science's. In Germany this granting of authority to science was not a concession to partial knowledge, rationalism or materialism. Perhaps the most powerful Nazi idea was the linking of scientific truth to eternal values. Consider Heidegger's conception of Greek science:

> For provided that the original Greek science is something great, the beginning of this greatness remains its greatest quality…. The beginning remains with us. It does not lie behind us in the past that is long gone, but is still before us. As the greatest, the beginning will outlast everything that is still to come. It will discard us as well. The beginning has to be penetrated to our own future; it is standing there over us as the distant decree, instructing us to recover its greatness once again. Only when we resolutely submit to this distant call to win back the greatness of this beginning, only then will science become the innermost necessity of our existence.[15]

Science must be grounded outside of time and space, as normally we conceive it, otherwise it would be only contingently or empirically true. The world was the way it was by nature, nature in its transcendental as well as temporal moment. It could be understood by a conception of science which respected its transcendent qualities.

Nazism's promise and purpose was to apply eternally valid scientifically demonstrable truths logically, ruthlessly and joyfully. The negative eugenics of Christian humanism would be replaced by the positive eugenics of National Socialism. Morality, so called, would be replaced by an appreciation of the "morality" of nature, of nature's dispensations, of nature's preferences for health, beauty, fertility, vigor, courage, and violence. This is the morality of the Life Philosophy. By living up to its laws, human beings would be able in time to transcend the limitations and compromises, and impurities of Time and Place and approach the Transcendent Truth. By using scientific truth, impure and imperfect human beings would become pure and godlike. Germans would become Aryans.

> An idea is not false or evil because the Nazis misused it…. Indeed, if we censored ideas that the Nazis abused, we would have to give up far more than the application of evolution and genetics to human behavior. And we would have to censor the study of evolution and genetics period. And we would have to suppress many other ideas that Hitler twisted into the foundations of Nazism. Stephen Pinker[16]

Some Concluding Remarks

In the face of the enormity of the Judeocide, concluding remarks always seem weak. Naturally, instinctively, one longs for justice to roll down like the waters, washing away sins of omission, commission, and human inadequacy in the face of implacable evil. Until that day or the End of Days, we will have to deal with events which burn away our complacency, hypocrisy and self-righteousness as best we can. This analysis indicates some steps we might take. First, while appreciating the power, progress and promise of science, we cannot substitute it for human judgments. No matter how scientifically convinced we are about a person or a disease or a predisposition, we must make a fully conscious and fully responsible choice about what should be done. Science, whether or not supported by social science, cannot and should not act as if it were an implacable deity or necessity. To be human is to make fateful, sometimes tragic, decisions, not to bow to a scientific or any other sort of Leviathan. We may have to act in partial knowledge. We may be metaphysically innocent. Whatever our true status may be, however we fit in the cosmos, we believe we are moral creatures, just as naturally as we possess our other attributes. Without responsibility, we free ourselves from meaning. If we remain cosmically insignificant, it is all the more imperative to act as if we live according to principles which enable us to seem meaningful if only to ourselves.

Our capacity to impute value to ourselves should not be taken to mean that we can construct the world any way we find convenient or meaningful. Although science can be misused or misapplied or misconceived, we must continue to apply its powerful methods to uncover or discover a reality we ignore only at our peril, physical and mental. We must continue to learn more about ourselves and our universe. Only then can we possibly live to our full capacities, as organic and moral beings. As Aristotle said, "Man is the animal who desires to know." In addition we must keep our concept and modes of understanding clear and distinct. I realize this sounds hopelessly academic. As academics what else can we do but our work as best we can, impotent as it may be in the face of the power state or irrational masses? For many reasons, some perhaps legitimate, German scientists did not do this. Nazi scientists did not want to do this. The confounding of science, social science, theology, morality in the name of public policy or politics in general is almost certain to work in favor of prejudice and bigotry and against all the civilizing forces in our societies. We may live in Spinoza's world of partial knowledge, but it is a far cry from celebrating a world of confusion, ignorance, willful or otherwise, and gratuitous violence. Acting responsibly with partial knowledge is the human condition. We can do it better if we keep our concepts and our modes of knowledge as clear and distinct as possible. For then they will be more resistant to demagogues from without and to our fears from within. We may not be able to resist the power state in its murderous activities. But we can resist the temptation to witness the annihilation of other human beings as some sort of ideal or sacrifice to the greater good or a false god. Or a true one, for that matter.

Notes

1.   Horowitz, I.L.: *Taking Lives,* New Brunswick, NJ, 1997, p.209. I thought of the title 'Taking Lives' before I began my systematic research, before, that is, I became aware of Professor Horowitz's book. There is more than coincidence in this—although what to call it escapes me. Professor Horowitz taught my first course in sociology, more years ago than either of us likes to remember. I must add that his chapter 13, Exclusivity and Inclusivity of Collective Death,' is among the very best treatments of this extremely difficult and emotional subject.

2.   Kershaw, Ian: *Hitler: 1889-1936: Hubris*, Norton, NY, 1998, p.xix

3.   Determinism became a pejorative term in the twentieth century and remains so for those who fear that, even as a metaphysical concept it will undermine human responsibility and

morality. It is the basis of one of the four fears which swirl around the concept of human nature. According to Pinker: 'If people are the products of biology, free will would be a myth and we could no longer hold people responsible for their actions.' Pinker, Stephen: *The Blank Slate: the Modern Denial of Human Nature,* Penguin, NY, 2002, p.139

4. Quoted by Pinker, *op. cit.*, p.56
5. Burleigh, Michael: *The Third Reich: a New History,* Macmillan, NY, 2000, p.13
6. Burleigh, *op. cit.*, p.99
7. Quoted by Pinker, *op. cit.*, p.26
8. Quoted by Burleigh, *op. cit.*, p.346
9. It is difficult to get a precise number of those killed under the policies of euthanasia. A reasonable estimate is about 100,000. According to Burleigh, 'Graphs showed how the deaths of 70,273 people [mostly mentally ill] achieved so far [1941] would have saved a projected 885,439,800 RM by 1951....' *Op. cit.*, p. 404. Note how easily the medical justifications become economic, especially in war.
10. [1933 Law for the Prevention of Hereditarily Diseased Progeny:] the law specified...congenital feeblemindedness, schizophrenia, manic-depressive illness, epilepsy, Huntington's Chorea, hereditary blindness and deafness, and severe physical malformation. Burleigh, *op. cit.*, p.354
11. Burleigh, *op. cit.*, p.254
12. Quoted by Farias, Victor: *Heidegger and Nazism,* edited by Joseph Margolis and Tom Rockmore, translated by Gabriel Ricci, Philadelphia, 1989, p.32. Heidegger here reflects a view held by German thinkers of virtually all political stripes since the French Revolution.
13. Burleigh, *op. cit.*, p.152
14. Burleigh, *op. cit.*, p.781. Harris was far from unique. Civilians were targeted by all the military forces of World War II.
15. Quoted in Farias, op. cit., p.101
16. Pinker, *op. cit.*, p.154

Do the Facts Matter? The Politicization of Science and the Betrayal of the American Trust

Gabriel R. Ricci

Since science as a profession and enterprise evolved from natural philosophy during and after the Enlightenment it has had a changing relationship with government and the administration of the state. When scientific investigation and experiment were first formalized within associations like the Royal Society, the Acadèmie des Sciences, and the American Philosophical Society in the United States, government had little to do with their ongoing operation. Over time, governments have routinely insinuated themselves into the workings of the scientific community. Between the two world wars, for example, totalitarian ideologies guided the practical outcomes of science, both Marxism and Aryan racial policy similarly dismissed the speculative aspects of science because they were either too bourgeois or too closely associated with Jewish scientists.

The use of the scientific community to produce a weapon of mass destruction during the Second World War forged a lasting relationship between government and science from which the two are not easily separated. And there are, of course, the notorious individual cases in which political authority has determined that certain scientific ideas must be suppressed in order to safeguard worldviews which supported institutional doctrine. Galileo's trouble with the Church and Robert Oppenheimer's falling out with the U.S. government are cases in point. Galileo exacerbated his problems by proving that he was as exegetically equipped as the next favorite Jesuit, and Oppenheimer fell victim to the Cold War paranoia; he did not help when he ventured to offer his position on foreign policy. The loyalty of a scientist at that time was very fragile and any echoes of leftist sympathies could undercut one's career. What really undermined Oppenheimer, since he had abandoned his socialist leanings long before, was mixing science and politics. His skepticism about the hydrogen bomb and the arms race which was surely to follow was not just a breach of loyalty, but a kind of trespassing. Oppenheimer had naively gone over the line and his punishment was a statement that science, if it were to remain vital, should restrict itself to practical and technological aspects of science and leave speculation about the future to policymakers. Scientists must be especially careful when they weigh in politically, but it is not the same for politicians who can liberally digress on scientific matters with little or no knowledge of the subject. At no other time in history has this double standard operated so blatantly than in the previous Republican administration in the United States.

The Manhattan Project saw an unprecedented alliance between science and government. Unwittingly it would also produce a succeeding epoch in which science would be at the heart of foreign policy. From the beginning, the new partnership was marked by unique tensions. Albert Einstein, a public pacifist at the time, would convey a message from the scientific community that encouraged the production of the atomic bomb. After the war, a similar ambivalence would dog Robert Oppenheimer, the scientific director of the project. Had he been as enthusiastic about the hydrogen bomb as Edward Teller, perhaps the government would have overlooked his foray into foreign policy. Unlike Teller, he was not a rabid anti-communist, so the dual role of scientist and foreign policy strategist was deemed untrustworthy and even subversive.[1]

Enlightenment science produced a special delineation between the antique world and so-called modernism. Following Bacon's methodological innovations, science was conceived as a means to control nature, as embodied in the famous dictum knowledge is power. In contrast, ancient natural philosophy was considered to be contemplative; Pythagoras's exaltation of *theoria* as an object of worship is the model for this worldview. Ironically the contemplative worldview sought political solutions and devised foundational theories of the state, while today science follows the money trail organized by state coffers. How did this transformation take place? How did science come to assume such a subservient role to the state when in its earliest incarnation it produced a metaphysically conceived, yet temporally grounded polity?

Recent trends in the United States suggest that politics has even hijacked the theoretical power of science. The problem partially originates in the way that ideology has usurped science and the articulation of facts, but the problem also comes from the way in which science has defined itself. In the first instance there is a clear violation of the Baconian inductive prescription for the facts to speak for themselves and not to simply fall in line with a preconceived theory. Thus we ought to be aware that while a theory might appear plausibly true, it should never be perceived as absolutely so. In other words, our theories, as Popper would later admonish, ought to be subject to being conclusively falsifiable. But science is also informed by Bacon's call for knowledge to be conceived as power. This instrumental function of science has been thoroughly appropriated by the rise of what John Kenneth Galbraith called technostructure or the institutional manner in which the technical ends of science become yet another way to achieve political goals. Within this structure science loses the independent status it has achieved within "the educational and scientific estate" and becomes subservient to political mechanisms.[2] The hard lesson that Oppenheimer learned when he became enmeshed in this newly forming institutional structure was that science was restricted to informing policy, its function was not to form policy. Galileo's predicament anticipated this tension when he instructed his detractors on the separate roles of defining how the heavens go and how one might get to heaven. The Church was adamant that they governed both spheres. Church doctrine, for example, could not tolerate bizarre notions like a multiple world theory. Their doctrinaire outlook would suffer irreparable damage; it was much easier to declare Giordano Bruno a heretic and burn him at the stake. The authoritative absorption of the technical prowess of science in both cases corrupt the ends that devolve from the momentum science produces on its own and allows authorities to determine ends that serve the arbitrary needs of transitory power structures. Needs that begin with maintaining political hegemony, or at least the appearance of national security as in the case of the strategies of the nuclear age. If the strategies that informed the nuclear arms race have taught us a lesson, it is that technical solutions cannot substitute for political ones. In the final analysis, what the use of reason and technology has done is to give the irrational (the prospect of mutual destruction) the appearance of being reasonable.[3] Science is far off its enlightenment course of freeing humans from the constraints and vicissitudes of nature if in its wake it has left us all facing residual pollutions and ecological insecurities, not to mention the threatening instabilities of a unfamiliar geopolitical landscape that was suppressed when international politics was technologically dominated by regulating the tensions between the two superpowers.

The Enlightenment has not gone without its critics. Hegel was the first to point out how the appeal of rationality produced a disenchantment of nature and a purging of the spiritual. Adorno and Horkheimer followed suit in their *The Dialectic of Enlightenment*, which viewed the Holocaust as the culmination of this technological domination of man and nature, all ironically in the name of making history or human interaction less irrational. Before scientific research was misdirected by politics, it had already been infiltrated by a competitive and Faustian spirit that overshadowed idealistic goals. (Goethe's Faust, before his encounter with Mephistopheles, despairs that his father's scientific apparatus was unable to penetrate Nature's secrets; what she is unwilling to reveal on her own one cannot pry loose with "thumbscrews, wheel or lever." In his moment of despair he bears the effrontery of a grinning skull and turns his attention to a potion whose essence is derived from slumber-bearing flowers [*Schlummersäfte*] Part I, 675-695). The prospects for any technology assessment that might oversee the state's role as both the producer and consumer of knowledge must first deal with the appeal of Bacon's identity of knowledge and power and its modern corollary: substituting technical solutions for political ones.

In many respects the administration of George W. Bush was a recapitulation of Ronald Reagan's administration, particularly in the way that industry and religion united to produce an autonomous brand of science. Presidents have had science advisory teams since the Eisenhower administration when Vannevar Bush was appointed the first presidential science advisor, but the collusion between industry and religion is a phenomenon that could not have been imagined with the first stirrings of the anti-scientific attitude and distrust of the intellectual elite in the early days of William Buckley's *National Review*. The *National Review* broadcast these views, but it was Buckley's *God and Man at Yale* that first began to question the cooperation between government and science that had prospered during the war. Buckley took aim at the social scientists of his day, but his criticism was intended for all scientists; he considered them to be much too critical of religion.[4] Buckley, perhaps unwittingly, set the stage for the more strident critique of science and scientists that would mark the Nixon administration and that would be resuscitated in the special merger between industry and religion in the Reagan administration and later in the second Bush administration.

Nixon did not devise the clever strategies that subvert conventional science; he was publicly angered when the community of scientists voiced their opposition to the Vietnam War. Scientists had had Nixon's ear for not only did he establish a full-bodied regulatory agency in the EPA, he also passed environmentally friendly legislation in the Endangered Species Act and the Clean Air Act. Nixon could not abide any criticism so scientists became persona non grata, and he soured on science. It was in 1972 that the EPA would ban DDT after the heroic efforts of Rachel Carson to reveal the devastating effects of its commercial use. This was the moment when industry took note of the way in which scientific discovery could be transformed into regulatory measures that would effect their bottom line. The subsequent smear campaign against Rachel Carson was not only driven by sexism; it also challenged the validity of her science. Carson was painted as an alarmist who subscribed to the cult of environmentalism, but industry was really agitated by the role that science began to take in informing laws designed to protect health and welfare and to prevent environmental degradation. Industry's attack against Rachel Carson ushered in the accusation of *scientific uncertainty* that would be bolstered and fortified during the Reagan administration. (There are still conservative, industry-friendly politicians who beat the drum against Rachel Carson; a Republican congressman from Oklahoma recently accused Carson of genocide for all the lives lost to malaria without the use of DDT. This charge, of course, ignores the fact that even mosquitoes eventually become immune to DDT applications, while it continues to do its destructive work across the food chain.)

The public display made by Newt Gingrich and his cohorts in 1994 to revive a conservative Republican agenda was accompanied by a systematic attack on the mainstream scientific community. Under Gingrich's inflated intellectual leadership an autonomous scientific program took shape, a position that first emerged under the auspices of Ronald Reagan who as governor of California encouraged a unique form of creationism, one that was informed by an ambiguous scientific approach. The result was scientific creationism or creationist science in which creative interpretations of Scripture were aligned with conventional science, like Darwinism. In another incarnation, creationism also tried to demonstrate the invalidity of evolution on scientific grounds. The political ambition in both cases was to introduce creationism alongside evolution in the curriculum. Reagan would take this new science to the White House; and, in addition, he would introduce industry friendly personnel into the EPA and the Department of Interior, two government offices dedicated to the protection and prudent use of the nation's natural resources. Under the leadership of Anne Gorsuch, the EPA would gain notoriety for drawing up a so-called "hit list" of scientists who were deemed unfriendly to industry.[5] Reagan's scientific team was questionable for all the foolish statements he made in public about acid rain at the time. They were all justified under the mantle of scientific uncertainty, just as scientific assertions about global warming face today. Still the Reagan use and abuse of science was not yet systemic; this would be left to the intellectual bravado of Gingrich and his fellow-travelers. Reagan, however, must be credited with the most concerted effort to turn a fantasy into a science with his Star Wars program. What we do see that would become even more troubling in the years ahead was the pressure placed upon science administrators to place an ideological spin on scientific information. In an effort to redirect and get control over the abortion debate, Dinesh D'Souza, a Reagan advisor, hatched a plot to use the public face of the surgeon general, who so successfully alerted the public to the dangers of smoking.[6] C. Everett Koop was pressured by D'Souza to produce a study suggesting that there were dangerous health consequences to abortion. Koop resisted and made it clear that there was no scientific data to support these conclusions. This was science getting the cart before the horse, an attitude that persisted into the second Bush administration where censorship and the debunking of science and scientists would be raised to a political art.

The practice of debunking science was cultivated and widely put into practice in the anti-regulatory campaign of the Gingrich Republican revolution. The Reagan administration revealed the problems associated with manipulating science advisors; Gingrich's crew would avoid this problem by pretending to be scientists themselves. Gingrich was a self-proclaimed futurologist, having been deeply moved by Alvin Toffler's work, and he liked to think that he had a grasp of the technological and scientific trends that would evolve over time. Before his political downfall, he publicly proclaimed that his sensitivity to technological and scientific discoveries gave him insights into social and cultural developments over centuries. He did not just envision a Republican revolution; he made pronouncements on cutting-edge science like nanotechnology after claiming to have studied at the country's most renowned science institutions. Informing public policy with the right sort of science had already been institutionalized in organizations like the Heritage Foundation and the George C. Marshall Institute, but Gingrich's intellectual audacity made him a one-man science and policy foundation.

The groundwork for Gingrich's scientific agenda had previously been established by Big Tobacco which had cultivated the technique known as "manufacturing uncertainty." This method is sanctioned by science itself which promotes a perennial skepticism in order to further the ends of inquiry at the heart of science. Big Tobacco had shown that instilling doubt was the most efficient way of producing controversy and challenging an accepted body of fact, as in the case of the risks of being exposed to second hand smoke. As far back as 1969 Brown & Williamson expressed their strategy: "Doubt is our product, since it is the best means of competing with the 'body of fact' that exists in the mind of the general public. It is also the means of establishing controversy."[7] The Advancement of Sound Science Coalition (TASSC), founded in 1993, would take up this banner. The group was heavily financed by Philip Morris. The group's mandate was to restrain politicians they deemed to abuse science. In the industry-friendly environment created by Gingrich et al. the banner of "sound science" led the way to regulatory reform. If regulatory reform could not be scuttled, the overbearing criteria for scientific review and the strict rules for risk assessment would at least slow down the process and wear down the opposition. At the time David Vladeck, a Georgetown University law professor described the political climate as "paralysis by analysis."[8] The strategy of adopting the call for "sound science" was cunning since it gave the appearance of taking the higher intellectual road and because it gave the appearance of an analytical and logical position, one that could only be accused of erring on the side of caution.

Industry did not have to rely on sympathetic legislators; there were eager enough consultants in Washington, D.C. to do their bidding. The long and illustrious career of James Tozzi is emblematic of the growing anti-regulatory ethos that challenged the validity of governmental science. Tozzi was a jazz musician manqué who had begun his career in the regulatory offices of the Corps of Engineers and progressively worked his way up to the Office of Management and Budget where he was able to make a close study of regulatory mechanisms and procedures. From the moment he left the Reagan administration, Tozzi actively worked on behalf of industrial entities whose bottom line could be affected by government science. The culmination of his career was shepherding the little known Data Quality Act in 2002. The brief two-sentence amendment to a budgetary measure provided industry with information about any study that might affect the bottom line. Under the guise of regulatory efficiency Tozzi provided industry with an inside track to challenging the validity of government science.[9]

The most notorious use and abuse of the facts in the second Bush administration must be the manipulation of evidence of WMDs in Iraq. This propaganda campaign has had disastrous consequences. The strategies employed in this campaign of misinformation are the same used against the scientific opposition to the Bush administration's manipulation of advisory and regulatory bodies mandated to guide national and international policy. Censorship and accusations of disloyalty became the primary tools employed against the opposition. Before her resignation in 2003, Christie Whitman had been accused of disloyalty for seeking "the facts." Anthony Zinni, a former commander-in-chief of Central Command, when he broke ranks with the Bush administration, was accused of treason. Part of being a wartime president meant creating a sensitive political atmosphere intolerant of dissent. This was the same environment the administration created for the scientific advisory community from its very beginning when it became obvious that George W. Bush was having difficulty appointing a lead scientific advisory team. It was well into the first year of the administration when John Marburger received the call to be the president's lead science advisor.

When the Union of Concerned Scientists produced a public statement on scientific integrity in the Bush administration in 2004, even Marburger, who had been in lockstep to Bush's control of scientific information, took note of the seriousness of this outcry. The federal government followed up that same year with legislation that forbade inquiry into a scientist's political background when they were being considered for advisory positions. The use of so-called litmus tests had been one of the administrations primary tactics in screening for loyal scientists; the other major strategy against the scientific community were well-placed gatekeepers intent on overriding any obstacles to the political goals articulated by the administration.

When he first came into office George W. Bush made quite a spectacle of the faith-based initiatives that were to substitute for meaningful domestic policy. This display made it quite clear that religion was going to be a major component of his administration. The fate of science would be reflected in the desultory way he appointed scientific advisors. While religion, in the form of creationism, did trump evolution, dictate stem-cell research, and inform birth control policy, it would not always be religion that blocked, censored, and obfuscated the facts. Industry-friendly advocates were well-schooled at this sort of sabotage. Before Jim Hansen accused the administration of censorship in 2004, the federal government had already (in 2002 the USDA established criteria for publication) established strictures that required scientists to receive permission to publish their scientific work and the two most regulation-conscious agencies in the government, the EPA and the Department of Interior, were led by enforcement officers like Jeffrey Holmstead (EPA) and Stephen Griles (Department of Interior) both of whom were adept at manipulating the data to satisfy ideological and industrial goals. Hansen's accusation that the Bush White House was deliberately interfering with scientists' communication with the public did not disclose the systemic institutional interference that had been going on for sometime. Hansen's case has come to the fore because of his admonitions about global warming, but the story of the chronic manipulation and censorship of the facts is told in the meddling and obstruction of ideological bureaucrats like Holmstead and Griles, both of whom cut their teeth with legal firms that represented the big industrial polluters. This experience put them in the ideal position to control scientific data. Holmstead repeatedly ignored the suggestions of his science advisors when they counseled him on the amount of retooling required in order for polluting industries to install state of the art, environmentally friendly technology. The figures were so skewed as to place extensive repairs and retooling beyond the necessary threshold. The Clear Skies Initiative, which Holmstead was invested in, would be guided not by science but by the market place. This policy was a subversion of the Clean Air Act whose standards Holmstead and others deemed to be a financial burden to industry. Having clean air would be determined by what was considered to be an optimal level of pollution across the board, and all the big polluters would participate in a system of trading pollution rights in order to maintain that level of optimal pollution. The policy, of course, ignored the fact that a really big polluter could buy up all available pollution rights and turn a region into an environmental nightmare. Similarly, Stephen Griles served in the Department of Interior as an industry advocate. Previous to government service he represented the mining industry and for all intents and purposes he did the same while serving as a deputy secretary at the Department of Interior. In this capacity Griles oversaw an Environmental Impact Study for strip mining. Typically an EIS would require a committee to investigate more environmentally benign ways of extracting coal from a mountain, other than removing their tops, deforesting the area, and producing toxic runoff. Under Griles's direction technical teams were told to remove language from their reports that might indicate a severe or negative environmental impact. Griles's real focus became expediting the permit process.[10]

Though John Marburger would call for a full investigation into the UCS's criticism of the Bush administration's politicization of science, he would ultimately dismiss their claims as grossly wrong on all counts.[11] In a pathetic attempt to defend Bush, Marburger strained to make it seem that George W. Bush agreed with the consensus view on global warming, when it was obvious to everyone that the administration would not admit to the role of human activity. Howard Gardner, one of Marburger's most outspoken critics, minced no words when he referred to Marburger as a prostitute on Diane Rehm's radio show.[12]

The Bush administration had multiple problems with facts. Whether it be in the form of concealing information, like the names of those attending Dick Cheney's notorious energy task force meeting in the early days of the administration, censoring important data from advisory committees or simply fabricating information, the Bush administration produced a separate reality that was sustained by friendly think tanks and monotonous talking points devoid of critical insight. The scientific community had to suffer the abuse of the Bush administration but so too did high ranking military officers, like Tony Zinni and Antonio Taguba, both of whom elected to reveal the facts. Taguba had the unenviable job of investigating what turned out to be the routine violation of the Geneva Convention and military protocol at Abu Ghraib. In all these cases the Bush administration waged a war on the truth, when it pretended to fight wars on more traditional fronts. The war on truth was a campaign against the facts. The Bush administration conflated having two sides to a story with having two sides to the facts, but its intellectual and psychological problems run deeper. The manner in which the Bush administration consistently held up and inflexibly defended questionable evidence; routinely manufactured apocalyptic scenarios on the basis of slim evidence; maintained false beliefs in spite of countervailing evidence; and denied evidence that undermines preconceived ideas suggests a psychological profile that Richard Hofstadter eloquently outlined in his seminal work, *The Paranoid Style in American Politics*. Hofstadter wrote at a time in which a paranoid mentality dominated the American political landscape, but he also took pains to historically review the various forms this syndrome has assumed, whether it be anti-Masonic, anti-Jesuit or in its anti-Semitic form. Whatever its incarnation, one can expect the same political pathology: the denial of compromise and mediation, a militant leader who sees things in terms of absolute good and absolute evil, and the demand that the evil enemy be totally eradicated. Partial success is construed as a sign of weakness and tantamount to defeat. Whatever form the enemy comes in, the paranoid personality perceives the enemy as having special powers to influence the mind: it may be foreign, religious demagogues, it can be the press, it could be elite Eastern intellectuals or even wayward scientists alarmed about global climate change. These powers, according to Hofstadter, are projections of the self and so the paranoid personality adopts the manners of its enemy in the way that the KKK modeled itself on Catholicism, down to the vestments, the rituals and the hierarchy. As the Bush administration faced off its scientific opposition it tried to outdo them in the "apparatus of scholarship" and in uncovering the facts. The result was a war on the truth, a war on science and a betrayal of the American trust.

Notes

1. In spite of protests from the scientific community Oppenheimer's security clearance was revoked by the Atomic Energy Commission in 1953. He had voiced his antagonism to the hydrogen bomb and his earlier foray into left-wing politics made him a target in the charged political atmosphere of the Cold War.
2. Salomon, Jean-Jacques, *Science and Politics*, translated by Noel Lindsay, p. 150 (The M.I.T. Press, 1973) quoted from Galbraith's *The New Industrial State* (Houghton Mifflin, 1967), p. 294.
3. Salomon, p. 241. While Enlightenment science promised the progressive amelioration of society, Salomon argues that there has been an anarchic escalation of the instrumental function of science that can no longer be reigned in by the power of knowledge itself. He compares this anarchic advance of technique and knowledge to the laissez-faire sensibility of capitalism in which we can expect corresponding injustices to prevail.
4. Mooney, Chris, *The Republican War on Science* (Basic Books, 2005), p. 29.
5. Mooney, p. 40.
6. Mooney, p. 46.
7. Mooney, p. 67.
8. Mooney, p. 70.
9. In a repeat of the Rachel Carson controversy, Tozzi worked with the Kansas Corn Growers Association and a group called the Atrazine Network to challenge a study by the EPA which called into question the use of the herbicide, atrazine, which was suspected of causing harm to the endocrine systems of frogs. See Mooney pp. 108-109.
10. Shulman, Seth, *Undermining Science, Suppression and Distortion in the Bush Administration* (University of California Press, 2006), p. 77. In a memo to the White House on Environmental Quality, Griles spelled out that a EIS draft should "focus on centralizing and streamlining coal-mining permitting" instead of thinking about the consequences of mountain removal.
11. Mooney, p. 232.
12. Mooney, p. 230. On March 4, 2004, Howard Gardner was a guest following an interview with Marburger.

Recommitting vs. Selling Out:
The Subtle Industrial Revolution among the
Amish of Lancaster County, Pennsylvania

Thomas R. Winpenny

While the Amish clearly desire to maintain their three-hundred-year tradition of cultural "separateness" or distance from the world around them, the burning reality for many in Lancaster County is that they live a "fish bowl" existence. Four and a half million tourists annually from metropolitan New York, Philadelphia, Baltimore, and Washington provide an abundance of observers—to say nothing of "English" neighbors, amateur and professional photographers, reporters for the electronic and print media, and area scholars. Implicitly or explicitly, each interloper has essentially the same agenda: identify and savor the uniqueness of the Plain People, and look for evidence of change in their patterns of behavior.

From the Broadway play "Plain and Fancy" in the 1950s to Harrison Ford's disguise as an Amishman in the film *Witness* in 1985, the Old Order Amish of Lancaster County have been hammered with unwanted attention and publicity. Furthermore, the Lancaster Amish are reputed to be the most modern and forward looking of the Plain People.[1] Thus, the inevitable question—are these deeply religious Anabaptists, with historic roots in Switzerland, adhering to their basic religious beliefs, or are they engaged in "selling out?" At the beginning of the twenty-first century this proves to be a fascinating question as there is abundant evidence to demonstrate that the Amish might appear to be doing both at the same time. Not surprisingly, local scholars have been busy proving both sides of the argument.

Parenthetically, the Plain People are living in a booming and growing county economy with a serious labor shortage. Agriculture remains the most prolific of any county in the Mid-Atlantic region with the value of product approaching three quarters of a billion dollars, industry is far more extensive than outsiders might suspect, and tourism continues to be a leader in the service sector that includes five colleges, five hospitals, one monster mall, and "outlet" malls. Almost 500,000 residents plus visitors cram on to antiquated roads engulfed in major reconstruction projects resulting in enough traffic problems to discourage some tour busses that hate to get ensnarled in traffic tie ups. Put another way, the romanticized Amish country of Harrison Ford's *Witness* continues to lose its innocence and isolation at a vertiginous rate.[2]

Perhaps the most predictable scholarship has been provided recently by Dirk W. Eitzen, professor of theatre, dance, and film at nearby Franklin and Marshall College who produced a film about the Amish first aired in November of 1997 that cleverly and subtly suggests they are "selling out." (It is possible that this film has "travelled" and has been shown on public television stations throughout the country beyond WITF in Hershey.) The documentary's thesis is quite simple: Plain People who hope to protect their most unusual style of life and intense faith can only do this through a dogged adherence to tilling the soil. What could be more basic and spiritually uplifting than perspiring and getting dirty walking behind a team of horses and a plow? By contrast, to turn to a variety of entrepreneurial endeavors or small shops or modest production facilities in order to reap cash is to abandon something sacred and to guarantee spiritual and cultural decline, or at least hypocrisy.[3]

Incidently, this shift in the allocation of capital and resources away from farming generally does not seem to require moving outside the Amish cultural context, and frequently not off the farm. For example, they are not purchasing Mc Donald's franchises or opening Buick dealerships nor are they putting money into NASDAQ (they would not invest in the stock market.) Normally, apart from the new microenterprises, their investment options have been limited to land, farms, apartments, and certificates of deposit from a bank.[4]

What are These New Entrepreneurial Endeavors?

Consider a hypothetical younger Amishman whose father is not in a position to provide him with a forty-acre farm on which to grow tobacco and support a family, particularly since farmland is disappearing and the price per acre increasing. (It is common knowledge today in Lancaster County that farmland might approach $10,000 per acre and that an Amishman might pay over $500,000 for a farm.) Given these harsh realities, it should be painfully obvious that not every Amish lad today can look forward to owning a farm. If it is also recognized that college is culturally out of the question, and that most people are not anxious to move to some other part of the country, then the remaining options are few. One ready alternative would be to open a harness shop on a corner of his father's farm. Here he would pursue a time-honored craft and would be meeting the needs of many within the Plain culture. Most of his business dealings are likely to be with other Amishmen. (The exception would be gentlemen farmers who maintain carriages.) He is no longer a farmer. Has he abandoned a vital tradition? Perhaps he has.

Consider Miller's Store just south of Leola that is tucked away on a farm and sells a wide variety of foods, with an emphasis on health foods. The store has been operating quietly for at least twenty years and seems to be known only to "locals." That is to say, when you drive in, you do not have to compete with tour buses for parking space. If and when Miller's decides to commit additional resources to advertising that might have gone into the farm, are they violating something sacred? Perhaps they are.

Consider a farmer in the Strasburg area who has run a greenhouse business on the side for many years. Of late he has put an addition on to his greenhouses to sell gardening tools and equipment. Worse yet, he offered the author a 2" x 4" calendar with the name of his business printed on it when I dropped in December to buy poinsettias. Has this man confused modern horticulture with traditional agriculture? Perhaps he has.

Here, then, are simply a few examples, perhaps relatively less offensive examples, of a trend the film professor saw as unfortunate. Obviously, the list could be expanded to include very serious production facilities that generate such items as bird houses, children's wagons, tool sheds, gazebos, mail boxes, buckets, baskets, baked goods, lawn ornaments, and furniture—frequently stacked neatly in substantial quantities along the roadside for all to see. These entrepreneurial efforts have the approval of the bishop, and they are required to remain relatively small family- run operations. (Only 14 percent had yearly sales in excess of $500,000.) At the same time, these enterprises generally represent aggressive wealth-gathering, and they surely manifest a strong dose of individualism, a quality recognized as a threat to the Plain People's sense of community. A name on a business card or on the front of a shop may seem insignificant or even obvious to the broader society, but in this distinctive culture it signifies a newly found individualism.[5] Indeed, there are novel dangers and temptations, including too much contact with the general public, that accompany these burgeoning activities. The film professor suggests these are potentially lethal—other scholars are less sure. Some suspect they are looking at the onset of the Amish Industrial Revolution.

Professors Kraybill and Nolt Analyze Amish Enterprizes in 1995

The definitive study of new Amish businesses by professors Donald Kraybill and Steven Nolt entitled *Amish Enterprise: From Plows to Profits*[6] reveals some of the unique characteristics of these endeavors. For example, they are exempt from the Social Security system except they must pay Social Security for any non-Amish employees and must pay the tax if they work for a non-Amish employer. OSHA has ruled that Plain People are exempt from wearing hard hats on a construction site so long as they wear a hat sanctioned by their religion. At the state level the Amish may stay outside the Workmans' Compensation system so long as they do not incorporate. In another state concern, Plain People establish small businesses with low overhead and use family members extensively, frequently pressing the limits of child labor laws. The existence of mutual aid programs eliminates the need to pay into health insurance and pension plans. They benefit from a large tourist market that comes seeking the Amish and the distinctiveness of their products. A strong work ethic is also part of the picture. Computers are frowned upon, at least in 1995, but a non-Amish partner might introduce them. Some utilize cash registers. There is no bankruptcy and very little failure. They do not "go to law" (file a law suit) with the aid of an attorney, but commonly submit to binding arbitration before a panel of their peers.[7] They might bring a lawyer to a zoning hearing. In many respects, then, the Amish bring a very beneficial third world cost structure to their enterprises. One unavoidable conclusion here is that these businesses thrive. Thus, while 65 percent of all small businesses in America fail within the first five years, no more than 5 percent of the Amish enterprises fail—a failure rate only 1/13th as high as the broader society's.[8] Or, as one Mennonite recently told this author about Amish shops: "There really are a lot of them, and almost all of them succeed."[9]

Parenthetically, as American business culture changes at a furious pace the new Amish entrepreneurs have no desire to be found wanting. Accordingly, the findings of Kraybill and Nolt from 1995 are currently being revised by Kraybill to include extensive use of computers.[10] (A recent conversation with a local accountant revealed that the Amish businessmen he associated with were generally equipped with laptop computers and cell phones.)[11]

The conclusion to the Kraybill and Nolt study cites one Amishman who had gained insights very comparable to those of Frederick Winslow Taylor—most assuredly without ever having read the writings of the father of Time and Motion Study. "One shop owner, writing about 'wasted motion,' calculated that if eight shop workers waste several seconds with each move, it adds up to $5,760 yearly loss for the owner."[12] Could it be that the Amish are careful with a dollar?

In the end, Kraybill and Nolt do not condemn. "Forced to leave the farms of their past, the Amish have refused to enter the alien culture of corporate America. Their embrace of microenterprises represents a tenuous cultural agreement—a midway point between their pastoral past and the world of high- tech industry."[13] The sympathetic authors further conclude: "They have struck a bargain that nourishes their economic health without conceding their cultural soul..."[14] Thus far, then, Kraybill and Nolt are as understanding in their views of the Plain Peoples' shops as the film professor's tone, if not language, is judgmental and condemning.

Conrad Kanagy Looks at the Buying and Selling of Farm Land

Another relevant voice in this debate is Conrad Kanagy, a professor of sociology and Mennonite pastor who has tracked the recent buying and selling of farmland through public records in the Lancaster County Courthouse. Armed with membership directories of many Amish, Mennonite, and Brethren religious groups, Kanagy has worked to determine just who is selling and who is buying farms throughout the county. In his initial efforts for the years 1984-1994 he uncovered some surprising results that his later research has not overturned. Essentially, Plain People—the Amish and the Weaverland and Groffdale Mennonites—had purchased a total of 154 farms while more modern Mennonites had sold twenty-five and the Brethren had sold thirteen. The "English" had purchased four while corporations (mostly commercial and residential real estate developers) had purchased twenty-six. Another 146 purchasers (about 40 percent) could not be identified, or at least associated with a particular religious group.[15]

Essentially, the Plain People were purchasing farmland while other members of Anabaptist religious groups were selling. Ironically perhaps, Kraybill believes that the emerging small shops have provided the wealth in many cases to buy the farms.[16] What is most critical in this farmland analysis is Professor Kanagy's observation that the Amish are buying land primarily out of religious conviction—because they want to offer their children a chance to be farmers. Put a bit differently, the half million dollar purchase prices do not necessarily constitute the very best economic investment. Indeed, the decisions might well constitute moderately irrational economic behavior. The irrationality of this decision simply reinforces the premise that it stems from religious persuasion and values.[17]

There are at least two additional thoughts that should be added to Kanagy's findings and analysis. First, when his research initially hit the local newspapers in 1995 the *Lancaster New Era* tried to sensationalize what was already a popular news story by talking about an "Amish land grab,"[18] thereby taking a trend that represented something virtuous and turning it into something less than virtuous. Second, there may be wishful thinking on the part of Plain parents who shifted emphasis from farming to microenterprises if they believe they can readily lure their children back into agriculture. There is at least abundant anecdotal evidence through the 1980s and 1990s in Lancaster County of parents (admittedly outside the Anabaptist tradition) trying to keep children in farming against their will, and having very little luck in doing so.[19]

Accordingly, the industrial revolution in microenterprises may not be reversible.

Resolving Conflicting Scholarly Views

As the Amish population continues to grow in Lancaster County, a plunge into microenterprises appears to be one reasonably good way for the Plain People to stay in south central Pennsylvania, maintain their communities, and survive economically. (Currently there are about 18,000 Amish and roughly 85 percent of the young can be expected to join the Church as adults and thereby continue as members of the community.[20]) Professor Eitzen, who brings a cynical eye and a mocking tone to the subject, sees the business alternative as the path to destruction—or if not destruction, at least hypocrisy. Professors Kraybill and Nolt, who are essentially insiders in the Anabaptist world, offer considerably more understanding by contending that the choices being made are entirely reasonable in light of the circumstances. Professor Kanagy contends that the Plain People of today should be commended for supporting their beliefs with their investment dollars in the purchase of farms.

Conclusion

The real issue here is whether new modes of wealth-gathering or more intensive wealth-gathering together with more contact with outsiders is likely to erode traditional values including religious commitment over time? The Biblical example of Abraham in the book of Genesis made it quite clear that riches need not destroy spiritual zeal. In more recent history, however, the evidence is more disturbing. The Puritans of early seventeenth-century Massachusetts Bay may have established their "Wilderness Zion" or "City Upon A Hill" for several decades or even several generations, but at the end of the century the merchants had become more influential than the clergy.[21] Or, as Cotton Mather put it, "Religion begot prosperity and the daughter devoured the mother."[22] In eighteenth century Philadelphia the Quakers readily sensed the tension between the Meeting House and Counting House and understood that the Counting House and material abundance would win out in the end.[23] Early in the nineteenth century, John Adams asked Thomas Jefferson: "Will you tell me how to present luxury from producing ... intoxication ... vice and folly?"[24] In the twentieth century, Joseph Schumpeter reminded everyone that "capitalism destroys the moral foundations upon which it is built."[25] Indeed, the accumulated historical evidence seems quite clear. In theory, everyone is potentially Abraham, but in reality most men and most societies find that over time their religious values and zeal are strenuously threatened by the distractions and temptations of wealth. Put another way, though no one can prove it today, the microenterprises have placed the Lancaster County Amish on a very slippery slope—even though they appear to be pursuing a reasonable course of action; even though they are remaining within their communities, and even though they have the approval of their bishops. This, then, could well be the beginning of an Industrial Revolution among the Plain People that, in time, will generate extraordinary and never-intended cultural change. It is relatively easy in the mind's eye to picture the transition from Amishman with hay rake to Amishman building gazebos. It is far more difficult to picture the transition from Amishman building gazebos to Amishman as totally assimilated middle-class suburbanite, but there is no guarantee that this could not take place over a few generations—as it already has for the Brethren and most Mennonites in the County.

Notes

1. The standard reference work on modern Amish life is by John A. Hostetler, *Amish Society*, 4th edition (Baltimore, MD: Johns Hopkins University Press, 1993). See also Walter M. Kollmorgen, *Culture of A Contemporary Rural Community: The Old Order Amish of Lancaster County, Pennsylvania* (Washington, DC: U.S. Department of Agriculture, 1942) and Donald B. Kraybill, *The Riddle of Amish Culture* (Baltimore, MD: Johns Hopkins, 1989). A new and rather sensational work by Jim Fisher, *Crimson Stain* (Baltimore, MD: Berkeley Books, 2000) examines the first Amishman to be convicted of homicide. Ed Gingrich of western Pennsylvania went out of his mind and later murdered his wife. The author contends that Gingrich was confused by evangelical missionaries who endeavored to save him from life in an Amish "cult."

2. Lancaster County consistently boasts of the lowest unemployment rate in the state of Pennsylvania. Currently, Acme Supermarkets warehouse cannot find unskilled workers to hire at $10.50 per hour. See Also, U.S. Bureau of the Census, *Twenty-Second Census, 2000. Agriculture.*

3. Dirk W. Eitzen, "The Amish and Us," WITF, Hershey, Pennsylvania, Public television documentary, aired November 5 and 9, 1997. Assisting Professor Eitzen were Franklin and Marshall professors David Tetzlaff, Simon Andrews, and Roger Thomas. Funding was provided by the Corporation for Public Broadcasting, Pennsylvania Humanities Council, Lancaster County Planning Commission, Franklin and Marshall College, and WITF. The real condemnation in the film is in the *tone* rather than the *language*—providing the producer with "plausible deniability" in the event he is accused of "Amish bashing." A colleague of the filmmakers who shall remain anonymous described the creation as "smart aleck, smart-assed, and self righteous." See also Franklin and Marshall press release dated October 8, 1997. The film is available from the station for $19.95. (1-800-242-0000).

4. See Hostetler.

5. See Donald B. Kraybill and Steven M. Nolt, *Amish Enterprise: From Plows to Profits* (Baltimore, MD: Johns Hopkins University Press, 1995). There is one allegedly unsavory business that has been growing for some time in the Plain community, and it has never been pleasant to even mention: the so-called "puppy mills." The Humane League of Lancaster County (a bastion of upper middle class propriety) alleges that many Plain farmers have turned to the mass production of dogs to offset a declining tobacco market. The League charges that these animals are essentially treated as commodities and then sold in the mass market as pets or in a highly specialized market for use in medical research. At present, Lancaster County has 231 licensed dog breeders—roughly twice the number of any other leading county in the state. The Plain farmers, by contrast, believe they are running a legitimate business and caring for the dogs in a responsible fashion.

    The August 14, 2000 issue of the *Lancaster Intelligencer* reported that thus far this year "seven new kennel proposals have gone before local municipalities, all submitted by Amish or Mennonite farmers." (pp. 1, 4) At the time the article appeared four had been approved, one denied, and two were pending. Two days later the two pending were rejected.

    In one recent "kennel dispute" the Pennsylvania Attorney General Mike Fisher instituted litigation against kennel owners Joyce and Raymond Stoltzfus of Puppy Love Kennels of Peach Bottom, Lancaster County and ultimately required them to return roughly $30,000 to dissatisfied customers. (pp. 1, 4) The dispute continued in the Lancaster newspapers through August of 2000.

Assuming the industry continues to grow, there will be ongoing controversy. Kraybill and Nolt cite the problem on pp. 206 and 207, but do not attempt to render a judgment.

Perhaps the only other "business" that competed with kennels for unsavoriness was the drug ring in Lancaster and Chester Counties that led to the indictment of eight members of the Pagan motorcycle gang and two Old Order Amishmen: Abner Stoltzfus and Abner King Stoltzfus in 1998. Of course this story was sensationalized in local headlines and elsewhere. See David Remnick's "Bad Seeds: In the Heartland of the Amish, the Outside World Finally Intrudes," *New Yorker*, July 20, 1998, 28-33.

6. Ibid.
7. Donald B. Kraybill, "Amish Enterprise: From Plows to Profits," presentation to the Elizabethtown College Faculty Forum, January 31, 1996—followed by a lengthy question and answer session. An editorial in the *Lancaster Intelligencer* for October 26, 2000 reports that "U.S. Rep. Joseph R. Pitts is trying again to push his bill to exempt Amish youths from some federal child labor laws." The story notes that, "The bill would allow children under 18 to work in Amish businesses such as saw mills and furniture factories where potentially hazardous machinery is used, but not to operate this machinery." For the second year this bill has passed the House and may be attached to a Senate appropriations bill. The editor opposed the bill as a "bad precedent that could open the door for others to seek special treatment based on their religion." On December 1, 2000 Representative Pitts brought Senator Arlen Specter to the Gap Fire Company in eastern Lancaster County to hear directly from Amish families who wanted their 14 to eighteen-year-olds to be permitted to sweep floors, sort wood, and run cash registers in saw mills. An Amish church leader contended, "This labor law conflicts with our religious beliefs."
    Lancaster New Era, November 30, 2000.
8. Ibid.
9. Conversation with Galen Sauder, Mennonite middle school teacher, in Lancaster County on July 17, 2000. One of the most far-reaching enterprises that might be considered Amish is actually run by a Mennonite, Chet Beiler, who utilizes Amish labor to construct the much-sought-after "Amish gazebos" for the California market. Conversation with John Leaman, January 4, 2001.
10. The Associated Press, "Amish Wrestle With Computers: PC's Becoming Necessary in Business As Plain People Leave Farms," *Lancaster Intelligencer*, July 31, 2000 cites Donald Kraybill who notes there are 1,500 Old Order Amish businesses in Lancaster County, 60 percent having been created since 1980. "This is the first generation ... very aggressively moving into business." They very clearly grasp the importance of computers in modern business—even though they believe that the Internet is full of "junk." The Amish entrepreneurs adopting PC's are living with some fear of a backlash from the bishops that could force them to choose between abandoning the computer or some form of punishment—essentially excommunication or shunning. Indeed, consider the case of Moses B. Smucker who was quoted in a rebellious mood in the July 31 story. He said: "I see a danger within our people of self-righteousness. Someone wants to be plainer than someone else. Plainer than you and proud of it." This tone was not lost on Smucker's bishop nor on his fellow Amishmen. Clearly, someone paid Moses a visit and reprimanded him for his brashness and no doubt reminded him of what would happen if he did not experience a change of heart and recant. Not surprisingly it followed that on August 4, 2000 Moses B. Smucker wrote a letter to the same newspaper and apologized for his words and attitude. "I am embarrassed and humiliated and I want to apologize to the Older Order Amish Mennonites, and to the general public. In the future I will try to be more careful with my words. I am ... working with the church to be in compliance with their rules and regulations. I am truly sorry for the people I have hurt..... I am truly sorry that this had to happen." *Lancaster Intelligencer*, August 4, 2000. This episode captures some of the anguish that one Amishman experienced

when tempted by alluring technological change—all the while claiming to be unimpressed with the alleged self-righteousness of his brethren. (One radical solution—perhaps totally unacceptable to Moses B. Smucker—might involve the embattled Amishman becoming a Mennonite and thereby gaining greater freedom.)

11. Conversation with Dale Weaver, accountant for Simon Lever, July 20, 2000.
12. Kraybill and Nolt, 258.
13. Ibid.
14. Ibid., 259.
15. Conrad Kanagy's research initially reached the public in an article by Ed Klimuska, "A Farm Surprise: Amish Are Buying Most Farmland Sold in County in Last 10 Years," *Lancaster New Era*, October 23, 1995, pp. 1 and 4. The story generated considerable public interest and debate.
16. Ibid., 4.
17. Ibid., The "irrationality" resides in purchasing more land when demand for (and thus production of) one of the Plain Peoples' leading crops is plummeting. For example, in 1992 some 9,860 acres of tobacco were harvested in Lancaster County generating 19,027,200 pounds of production, but by 1997 only 6,800 acres were harvested generating only 14,310,000 pounds of production. For the same period, Lancaster County's percentage of Pennsylvania's tobacco crop dropped from 91.3 percent to 83.9 percent. See *Annual Report, 1998-1999 Statistical Summary* from the Pennsylvania Department of Agriculture, Harrisburg, Pennsylvania. Reports published in the *Lancaster Intelligencer* for January of 2001 suggest a weak tobacco market and low prices. One ray of hope resides in the possibility of higher prices in the future for those who get involved in growing nicotine-free tobacco. By February of 2001 local newspapers were reporting that area farmers were committing to growing nicotine-free tobacco.
18. Ibid., 1.
19. The author has personal knowledge of three of these cases, one of which ended in a great tragedy, indeed a multiple homicide. Many rural youth in Lancaster County today, though not the Amish who do not attend high school, are quite likely to attend a suburban high school where the greater majority of the young people are never asked to perform farm labor. The farm youth, of course, recognize the difference immediately and resentment frequently builds.
20. Kraybill and Nolt, pp. 24 and 29. Professor Kanagy recently noted that the "Amish population is doubling every twenty years." He also observed that out of roughly 100 Amish weddings that take place in Lancaster County each year, "only 10 to 15 couples will work in farming." See *The Etownian* (Elizabethtown College newspaper) September 22, 2000, 10.

   In the face of all this change, according to Kraybill, a higher percentage of young people left the Amish faith in the 1950s and 1960s than are leaving today. Perhaps the Church today is more understanding.
21. See Edmund Morgan, *The Puritan Dilemma* (Boston: Little, Brown, 1958) or Bernard Bailyn, *The New England Merchants in the 17th Century* (New York: Harper and Row, 1955), or any of a host of works by the venerable Perry Miller.
22. Stephen Innes, *Creating the Commonwealth: The Economic Culture of Puritan New England* (New York: W.W. Norton, 1995), 26.
23. Frederick B. Tolles, *Meeting House and Counting House* (New York: W. W. Norton, 1963).
24. Innes, 26.
25. Ibid.

Technology, Tribes, and Environmental Racism: From Techno-Oppression to Tribal Sovereignty

Kyle Powys Whyte (Citizen Potawatomi)

Introduction

Indian tribes in North America have been deeply affected by risky technologies, from large-scale dam projects to hazardous waste landfills. "Techno-oppression" refers to the process whereby dominant nations like the U.S. sanction the implementation of risky technologies on or near lands depended on by tribes and other vulnerable communities. The sanctioning of risky technologies creates an oppressive decision scenario because (1) direct resistance on the part of tribal members forces them to accept the costs and hardships associated with civil disobedience, legal action, protest, community organization, and political unrest and (2) living with the risks engenders uncertainty about whether "tribal traditional lifeways" can be maintained. "Tribal traditional lifeways" refer to the activities, human networks, expertises, and environmental conditions that are necessary for the tribal members to flourish (NEJAC).

The challenges of techno-oppression are as pressing today as they have been at any time during the colonization of North America. Tribal lands continue to be targeted for risky technologies, like the nuclear waste storage facility proposed for the Yucca Mountain area, which is part of the Western Shoshone Nation homelands (Walker and U.S. Nuclear Regulatory Commission.). Tribal lands are being affected by climate change, raising difficult issues insofar as the responsible technologies may be thousands of miles away and impossible to isolate (Singer). Tribes also face decision scenarios about what energy technologies they should accept, from coal to alternative energies and green technologies (Grijalva).

I will begin in section 1 by defining "techno-oppression" and two specific kinds of environmental racism inflicted under it in the tribal context, which I refer to as "explicit" and "subtle" environmental racism. What remains unclear is whether there are viable political approaches today for eliminating techno-oppression. In section 2, I give a case overview of one political approach—"treatment as state status" (TAS status). I will argue in section 3 that TAS status commits an error that should be avoided at all costs in the tribal context. TAS status is based on a theory of tribal sovereignty that does not eliminate environmental racism, despite its relevance to breaking what are often oppressive decision scenarios. It fails to eliminate environmental racism because it does not respect the tribal perspectives on environmental risk and its management. My goal in defending this argument is to suggest that philosophers of science and technology and "science and technology studies" practitioners should turn some of their attention toward exploring the possibility of a theory of tribal sovereignty that (1) values tribal interpretations of technology risk and its management and (2) remains binding on state and federal governments.

Techno-Oppression and Technology Transfer

I will begin by defining "techno-oppression" and the kinds of environmental racism associated with it in the tribal context. The literature on tribes and risky technologies could be said to fit into a larger body of literature on the general phenomena of "technology transfer." "Technology transfer" is generally used in the context of international economic development and refers to the process whereby wealthy nations export their technologies to poor nations that do not share the same technical heritage and infrastructure (Tiles and Oberdiek, 138-39). One issue in "technology transfer" is that the exported technologies do not have the same impacts in the host nation as they do, or did, in the exporting nation.[1] Usually, the host nation experiences negative impacts, which means that technology transfer can impede development.

Whether technology transfer always produces negative consequences is highly contested in both academic and policy circles (Borgmann; Cohen; Tiles and Oberdiek). The "Green Revolution" is an example of technology transfer the impacts of which are hotly debated in cases where the agricultural techniques were exported to poor nations in Asia and Africa (Shiva). Some argue that the new agricultural techniques were exploitative and ruined the conditions of agriculture in the countries that adopted the new methods; others argue that the new methods have created conditions for prosperity that previously would not have been possible (Tiles and Oberdiek). Other cases of technology transfer exemplify a non-profit drive to empower people though technologies previously unavailable to them. The Grameen phone is intended to create conditions for grassroots economic development by the introduction of cell phones into Bangladesh (Sullivan). Of course, there are also debates about the success of the Grameen phone (Selinger).

"Techno-oppression" can be understood as a version of technology transfer where a host nation, community, or tribe does not consent to a large-scale technology being exported and implemented on or near their lands. In addition, the values of social and economic development reflected by oppressive technologies differ drastically from the values of the host nation, community, or tribe. I am concerned in particular with the tribal context of techno-oppression. Consider when the Dalles Dam was built and opened in 1957 in the middle of the Columbia River. The tribes living on or near the river did not have any choice as to whether the technology (dam capabilities) would be exported and implemented in their homelands. The dam reflected industrial "Cold War" development values that contrasted sharply with the tribes" traditional lifeways. The dam's reservoir flooded the tribal fisheries, destroying the Indian communities and cutting off their sources of livelihood (Barber).

A growing literature reveals how common events like what happened at Celilo Falls are in Indian country. (Grinde and Johansen; LaDuke; Weaver *Defending Mother Earth:Native American Perspectives on Environmental Justice*; Shrader-Frechette). The cases range from Navajo communities that are not sufficiently protected from the environmental and workplace hazards caused by uranium mining (Grinde and Johansen) to the pollution caused by retired industrial facilities on the St. Lawrence River that affected Akwesasne Mohawks (Tarbell and Arquette; Arquette et al.).

Examples like this one properly called "oppressive" because of the decision scenario that arises vis-à-vis risky technologies. Grace Thorpe writes that,

> The Great Spirit instructed us that, as Native people, we have a consecrated bond with our Mother Earth. We have a sacred obligation to our fellow creatures that live upon it. For this reason it is both painful and disturbing that the United States government and the nuclear power industry seem intent on forever ruining some of the little land we have remaining. The nuclear waste issue is causing American Indians to make serious, possibly even genocidal, decisions concerning the environment and the future of our peoples. (Thorpe 49)

Thorpe's point is that the imposition of the technologies forces tribes into "lose-lose" decision scenarios. The decision scenario referred to by Thorpe is one where there really are two choice options. Grace Thorpe writes that, "The Great Spirit instructed us that, as Native people, we have a consecrated bond with our Mother Earth. We have a sacred obligation to our fellow creatures that live upon it. For this reason it is both painful and disturbing that the United States government and the nuclear power industry seem intent on forever ruining some of the little land we have remaining. The nuclear waste issue is causing American Indians to make serious, possibly even genocidal, decisions concerning the environment and the future of our peoples" (Thorpe, 49). Thorpe's point is that the imposition of the technologies forces tribes into "lose-lose" decision scenarios. By "decision scenario," I refer to the scheme of practical choices that are available to tribal members and leaders vis-à-vis risky technologies like large-scale dams, uranium mines, nuclear waste storage facilities, and so on. Following Marilyn Frye, oppression can be understood as decision scenarios where every available choice option entails a significant and unwanted consequence (Frye). In a more specific sense, "techno-oppression" refers to the process whereby dominant nations sanction the implementation of large-scale technologies, like waste incinerators and uranium mines, on or near the lands of vulnerable communities or tribes, the result of which is that tribes are forced into techno-oppressive decision scenarios. The oppressiveness of the decision scenario is what makes techno-oppression a specific version of technology transfer.[2]

I do not aim to prove to what degree and extent tribes in North America face techno-oppression today. By defining techno-oppression, I am simply pointing out a way of understanding how imposed large-scale technologies create decision scenarios that are oppressive on those who have to live through them. All cases of techno-oppression are unacceptable and measures should be taken to eliminate them. In the tribal context, techno-oppression is almost always accompanied by the infliction of environmental racism on specific tribes and Indians in general. I will describe two kinds of environmental racism in order to more completely contextualize the impacts of techno-oppression on tribes.

The literature suggests two forms of environmental racism inflicted under techno-oppression in the tribal context. The first kind is "explicit" environmental racism, which occurs when one group forcibly imposes its values on another group without availing to any relevant reason (Bullard; Westra and Lawson). Explicit environmental racism occurred in the Celilo Falls case where the industrial, "Cold War" values behind the dam projects were privileged by the dominant nation sanctioning them. Native values were considered to be expendable and could be sacrificed for the interests of U.S. expansion and power (Barber).

Explicit environmental racism puts a lot of stress on its victims. Winona LaDuke writes about this stress in relation to the Northern Cheyenne tribe's struggle with coal mining.

> The [coal] mining [development] would affect a nation of people, with some 3,000 residents on their pristine reservation. The influx of workers, machinery, and infrastructure would impact the community socially, politically, and culturally. Sociologists refer to the ramifications of such development as "boom town syndrome." It is not considered to be a healthy environment for the host population and is exacerbated when the local host community is a different color, race, and culture from the newcomers. (LaDuke 84)

The development of coal mining on or near the lands belonging to a tribe imposes non-tribal values on the tribal members as a matter of explicit environmental racism. This can result in what Agnes Williams calls "ethnostress" or "… what you feel when you wake up in the morning and you are still Indian, and you still have to deal with stuff about being Indian—poverty, racism, death, the government, and strip-mining. You can't just hit the tennis courts, have lunch, and forget about it, you will still have to go home" (LaDuke 90). Moreover, LaDuke writes, "Ethnostress comes from the structural issues that affect every reservation.… Most statistics place reservations in the third world economically, something that the energy and mining corporations understood when they negotiated leases on Native lands. Land- and natural resource rich, the Northern Cheyenne continue to live in immense material poverty" (LaDuke 90). Explicit environmental racism results in clear burdens being placed on the shoulders of tribal community members, triggering ethnostress. The truly "choiceless" decision situation imposed by techno-oppression is conducive to ethnostress and harmful to tribal members in many ways.

The second form of environmental racism is "subtle" and refers to the collective forgetting and misrepresentation by those who are part of the public of the wealthy or dominant nation that imposes techno-oppression on tribes. Jace Weaver writes,

> [Signs of environmental devastation] are evident in proposals that would turn Indian lands into dumps for toxic and nuclear wastes and in the NAFTA-engineered collapse of the Mayan corn-based economy. It is, however, one of the ironies of the colonial relationship that even as Native lands and peoples are destroyed, they are also idolized and idealized. Non-Natives in the Americas have always exhibited a strange attraction-repulsion relationship toward the indigenes of this hemisphere.… In the U.S., Oklahoma can trumpet that "Oklahoma is Native America," while the Indians whose heritage it celebrates are discriminated against and marginalized. (Weaver "Notes from a Miner's Canary" 3)

"Subtle" environmental racism is not the same for every group. (Westra and Lawson). In the tribal context, the subtle environmental racism is related to how the struggles of tribes are erased from the minds of many U.S. citizens. Moreover, Indian culture is often portrayed romantically at the expense of understanding what is actually going on in Indian Country—especially cases of techno-oppression (Sturgeon; Harkin and Lewis; Krech). Techno-oppression creates conditions for the possibility of subtle environmental racism insofar as the available choices do not provide the opportunity to voice tribal concerns to the U.S. public at large. Having to deal with technology risks with limited resources makes doing anything else practically impossible. In addition, many tribal lands are singled out due to the fact that they are remote and will not be an issue in the public sphere.

In the tribal context, techno-oppression is usually associated with explicit and subtle environmental racism. When explicit, the environmental racism has to do with the privileging of dominant values over Native values; when subtle, it has to do with the collective forgetting and misrepresentation of Native values in the dominant nation's public sphere. The point I would like to make is that resolving techno-oppression in the tribal context *also requires* resolving the two kinds of environmental racism. If it is possible to break the oppressive decision scenario, the method and means must also avoid environmental racism. In what follows in section 2 I will consider an example of a political approach to resolving techno-oppression having to do with water quality: TAS status. In section 3 I will evaluate TAS status, arguing that it offers the possibility of breaking techno-oppression but does not eliminate environmental racism.

Case Overview: Water Quality Standards and the Isleta Pueblo

The following is a basic case overview of an environmental conflict over water quality and an ensuing court decision. The goal of the overview is to illuminate a proposed political approach that could be a solution for eliminating techno-oppression, TAS status.

The Isleta Pueblo is a small tribe located in the state of New Mexico. A segment of the Rio Grande river runs through the tribal jurisdiction. Some of the tribal traditional lifeways, from fishing to agriculture to religion to recreation, depend on an optimum amount of water running through the tribal jurisdiction. A controversial issue in Indian Country is how tribes should be able to govern the water running through their jurisdiction in relation to U.S. state and federal government controls. Every tribe with a reservation has some reserved water rights that are clarified in part by what is referred to as the "Winters doctrine." which arose from the case *Winters v. the United States* in 1908. The case was brought by the U.S. to protect the parts of the Milk River flowing through the Fort Belknap Reservation in Montana against upstream diversions of water by non-tribal actors (Royster "A Primer on Indian Water Rights: More Questions than Answers"). The Supreme Court held that the tribes held a paramount right to the water flowing through the reservation and that this right was based on the fact that the purpose of the reservation system was to convert tribes to agricultural lifestyles, which required water in the arid West (Royster "A Primer on Indian Water Rights: More Questions than Answers," 65). This, of course, is ironic given that the intent to convert tribes to an agricultural lifestyle was among the major forms of colonial racism in North American history. On the whole, reserved rights are not entirely based on the agricultural purpose of reservations. Reserved rights can also be for the maintenance of fisheries, tribal traditional lifeways, and other purposes associated with having a homeland (Royster "A Primer on Indian Water Rights: More Questions than Answers").

Of course, having reserved rights does not mean that tribes have the freedom to exercise them in their relations between U.S. states and the federal government. Many environmental statues in the 1970s omitted tribes, the Clean Water Act being no different. It became unclear what tribes should do to take an active role in regulating pollution problems within their jurisdiction. The 1983 "American Indian Policy" directed federal agencies to encourage tribal self-government and the EPA became the first agency to follow suit. "The policy recognized "tribal governments as the primary parties for setting standards, making environmental policy decisions, and managing programs…consistent with Agency standards and regulations"(Sanders).

In 1987, amendments to the Clean Water Act (CWA) created a program for Indian tribes to be treated as states, which is referred to as treatment as state (TAS) status. TAS status allows tribes to determine certain regulations subject to the approval of the Environmental Protection Agency (EPA) (Royster "Environmental Federalism and the Third Sovereign: Limits on State Authority to Regulate Water Quality in Indian Country"). First, TAS status permits tribes to administer EPA's National Pollution Discharge Elimination System (NPDES) program, which serves to limit the discharge of pollutants from point sources (e.g., pipes, ditches, conduits, wells, containers, vessels) by establishing technology standards. Second, TAS status permits tribes to set their own water quality standards (WQS) for waterways within tribal jurisdiction. Any standards set by a tribe have to be at least as stringent as EPA's national WQS (Royster "Environmental Federalism and the Third Sovereign: Limits on State Authority to Regbulate Water Quality in Indian Country").

To qualify for TAS status, a tribe must meet the following criteria:

1. The tribe must be federally recognized and must be "exercising governmental authority over a Federal Indian reservation.
2. The tribe must have a governing body carrying out substantial governmental duties and powers.
3. The functions exercised by the Indian tribe must pertain to the management and protection of water resources which are held by an Indian tribe, held by the United States in trust for Indians, held by a member of an Indian tribe if such property interest is subject to a trust restriction on alienation, or otherwise within the borders of an Indian reservation; and
4. The Indian tribe must be "reasonably" capable, in the Administrator's judgment, of carrying out the functions. (Sanders 6-7)

The result of these criteria is that approval applications have to provide descriptions of the form and functions of the tribal government, including the tribal government's source of authority for carrying out these functions. A tribe also has to describe its authority to regulate water quality, including a map or legal description of the area over which it asserts such authority, a statement from the tribe's legal counsel (or equivalent official) which describes the basis for the assertion of authority, and an identification of the surface waters for which the tribe proposes to establish WQS. The application also requires that a narrative statement detailing the capability of the tribe to administer an effective WQS program be submitted. This statement must include a description of the tribe's previous management experience; a list of existing environmental or public health programs administered by the tribe and copies of related tribal laws, policies, and regulations, a description of the entity (or entities) which exercise the executive, legislative, and judicial functions of the tribal government; and a description of the technical and administrative capabilities of the staff to administer and manage an effective WQS program, or a plan which proposes how the tribe will acquire additional administrative and technical expertise. The plan must also address how the tribe will "obtain the funds to acquire the administrative and technical expertise" (Sanders 7).

If approved for TAS status, tribal WQS and NPDES are enforceable by EPA and are commensurate to those set by states. This means that if the technology standards issued by the state permit discharges that violate tribal WQS, then, owing to the latter's TAS status, the state will have to modify its technology standards and ensure that the point source facility complies with the tribal WQS.

The Isleta Pueblo tribe created WQS for their segment of the Rio Grande River that were approved by EPA in 1992. The tribe's WQS were more stringent than New Mexico's technology standards for a wastewater treatment plant that discharges treated water five miles upstream from the reservation. One of the reasons why the tribe promulgated more stringent WQS was that one of the designated uses of the river is primary contact ceremonial use. Other uses include warm water fishery use, primary contact recreational use, agricultural water supply use, industrial water supply use, and wildlife usage (Isleta). One of the pollution discharges from the wastewater facility was arsenic and, given these designed uses, it is easy to see why the tribal members were concerned with water quality.

To comply with the tribal WQS, the city of Albuquerque would have to pay millions of dollars in compliance costs. To avoid paying the costs, the city sued the EPA, alleging that the agency's approval of the Isleta WQS was invalid, which resulted in *Albuquerque v. Browner* (1996). The federal court held that EPA validly required the upstream wastewater facility to comply with the tribal WQS and modify its facility to meet the standards. The parties reached agreement on a revised NPDES permit for the treatment plant (Royster "Environmental Federalism and the Third Sovereign: Limits on State Authority to Regbulate Water Quality in Indian Country," 20).

The significance of *Albuquerque v. Browner* is debated from many different perspectives (Sanders). I want to focus on the idea that the tribe's victory turned on their being approved for TAS status and having EPA sanctioned WQS. Without TAS status, the tribe would have had no good options for coping with the arsenic risks. The tribe did not choose that the wastewater facility be placed upstream from it; and insofar as the facility met the state's standards, it would have been costly for the tribe to fight those being able to have its own WQS that were enforceable by EPA. While this case may seem to be milder than the Dalles Dam or Navajo uranium mining cases, it is still a case of techno-oppression insofar as the location of the facility was not tribally consented to and, without the possibility of TAS status, the tribe would not have had good options for protecting their tribal traditional lifeways.

Tribal Sovereignty and Technology Risks

The Isleta Pueblo WQS overview reveals an underlying techno-oppression: a wastewater facility was sending arsenic downstream, threatening tribal traditional lifeways and leaving tribal members with an oppressive decision scenario. The issue that I will explore is whether TAS status is an adequate solution for techno-oppression in the tribal context. I will argue that TAS status is inadequate because it based on a theory of tribal sovereignty that allows EPA to retain powers over what criteria are relevant for approving TAS applications, interpreting technology risks, and setting WQS. Retaining these privileges does not avoid environmental racism in the tribal context. The goal of my argument is to demonstrate the need for a theory of tribal sovereignty in cases of techno-oppression to the attention of philosophers of science and technology and science and technology studies practitioners. This goal opens a domain of scholarship that seeks to formulate a theory of tribal sovereignty that is acceptable to tribes in cases of techno-oppression.

We can begin to evaluate TAS status by asking two questions. First, does EPA's enforcement of tribal WQS guarantee sufficient powers to break techno-oppressive decision scenarios? Second, is the procedure through which tribes are approved for TAS status consistent with tribal sovereignty in terms of how technology risks and the tribe's ability to manage them are interpreted?

The rationale behind asking the first question is that an initial level of evaluation should check whether TAS status at least breaks the techno-oppressive decision situation. The answer to the first question is likely an affirmative. Previously, the tribe could have either lived with the risks or tried to resist the presence of the facility itself. TAS status splits the horns of the dilemma by making a third option possible, that is, the option of tribal WQS that are binding on other entities and enforceable by EPA. In this particular case, the Isleta Pueblo tribe has more options than it would have had.

An affirmative answer to this first question only establishes that TAS status is "extensionally equivalent" with inherent sovereignty in terms of the enforcement of rights, standards, and other powers. "Inherent sovereignty" refers to the tribe's self-governance since time immemorial. "Extensional equivalence" refers to the idea that two theories might be different but that they make the same recommendations or prescribe the same policies or actions. TAS status is certainly *not* inherent sovereignty; however, as far as enforcement of WQS is concerned, it is extensionally equivalent. That is to say, both TAS status and inherent sovereignty would guarantee enforcement of powers like WQS.

In terms of techno-oppression, it is questionable whether the extensional equivalence of TAS status to inherent sovereignty is an adequate solution. One reason why it can be considered questionable is that enforceability only applies to WQS that are *already* promulgated by the tribe. This reason leads to the second question that I posed earlier in this section: Is the procedure through which tribes are approved for TAS status consistent with tribal sovereignty in terms of how technology risks are interpreted and managed?

The second question is harder to answer because it addresses the issue of whether the approval procedure reflects the tribes terms and not just EPA's terms. If TAS status and inherent sovereignty were relevantly similar in this respect, then the terms amenable to tribal sovereignty would permeate the procedural requirements. Darren Ranco and Ann Fleder write that,

> ...viewing [*Albuquerque v. Browner* (*Browner*)] as a promotion of tribal sovereignty, without acknowledging the precariousness of a situation where tribes associate themselves with the federal government by accepting TAS status and agency control, may be naive and dangerous... If tribes are willing to accept incremental steps of success towards an ultimate goal of self-government—self-government that is inherently limited due to its close ties to the federal government—then *Browner* is certainly a successful move in that direction. However, if tribes have a broader vision of self-government that offers them full autonomy and independence in their decision making, free from the federal government's control, then Browner represents an erosion of sovereignty for tribes. It entrenches tribes in a restrictive system, rather than promoting tribal sovereignty and culture without any guiding or limiting forces. (Ranco and Fleder 43)

Ranco and Fleder's key point is that the process whereby the Isleta Pueblo was approved for TAS status is entirely conceived of from the values of EPA. There was very little tribal input on how tribes should be evaluated as being capable of managing risks, how risks should be interpreted, what scientific criteria WQS should meet, and so on. Just to get TAS approval and, subsequently, have approved WQS, requires that the values of inherent sovereignty are suspended in favor of EPA values.

This is problematic in terms of what I discussed earlier regarding environmental racism. Explicit environmental racism occurs when values of one group are forcibly imposed on another group. If TAS status is only possible within the right restrictions set by EPA, then is techno-oppression avoided? My claim is that it is not insofar as the procedure that leads to the tribe's ability to actualize TAS status has the bad consequence of having to submit to EPA's values on technology risks and the tribe's ability to manage them. As Ranco and Fleder suggest, TAS status is not the promotion of inherent sovereignty, but a restriction on it, despite the fact that approved WQS are supposed to be enforceable.[3] It can also be said that TAS status does not avoid subtle environmental racism as well. As I stated earlier, one of the problems of techno-oppression for tribes is that the public of the dominant nation misrepresents tribal struggles. Requiring EPA approval on EPA's terms masks the compromises and struggles that tribes have to make to protect their homelands. It creates conditions whereby to become a state actor a tribe has to fashion itself according to EPA's conception of how a state actor should be structured.

The underlying problem is that TAS status is not based on a theory of tribal sovereignty that respects tribal interpretations of technology risks and how to manage them, which suggests that it does not avoid environmental racism. But, as in *Albuquerque v. Browner*, it did have binding power and enforceability. I make my argument in light of new research on how vulnerable communities, like tribes, can break techno-oppressive decision scenarios. There has been a lot of recent work on this in the philosophies of science and technology and science and technology studies (Dietz; NEJAC). There is also more work on environmental justice that raises important questions about how tribes should interact with federal agencies and states in ways that respect sovereignty (Mutz, Bryner, and Kenney). Work of this kind can only be relevant to tribes, however, if, in addition to offering recommendations that enforce powers capable of breaking techno-oppression through political approaches like TAS status, solutions are offered that are consistent with inherent sovereignty, which includes technology risks.

To put it more succinctly, solutions to techno-oppression should always include a theory of tribal sovereignty that begins with procedures which determine how technology risks should be measured, and includes the ability of a tribe to articulate and manage environmental standards and which provides independent means of enforceability. Above all else, the theory of tribal sovereignty has to be understandable to and binding on state and federal governments. Based on this demand alone, I think that we can anticipate even more efforts to resolve the pitfalls of techno-oppression. Taiake Alfred asks the question, "What will Native governance systems look like after self-government is achieved?" (Alfred, 26). This is the fundamental question and should guide our thinking. We should not ask what adequate solutions would look like in terms of the extensional equivalence of tribal powers, the ones they are supposed to have. Rather, we should ask what solutions would look like in a world where tribes have their inherent sovereignty respected.

Notes

1. This also applies to cases where a development technology was never used in the exporting nation but was intended to have a particular impact in the host nation.
2. Many cases of technology transfer do involve the consent of the members of the host nation, community, or tribe. Or the technology was adapted in conjunction with them . Oppression means something very specific and it is not applicable to all cases of technology transfer.

3.   See Ranco and Fleder for more issues regarding the relationship between tribes and EPA. In this essay I barely scratch the surface of many of the problems.

Bibliography

"EPA Policy for the Administration of Environmental Programs on Indian Reservations 2". November 8, 1984. <http://www.epa.gov/tribalportal/basicinfo/epa-policies.htm>.

Alfred, Gerald R. *Peace, Power, Righteousness: An Indigenous Manifesto*. Don Mills, Ont.: Oxford University Press, 1999. Print.

Arquette, Mary, et al. "Holistic Risk-Based Environmental Decision-making: A Native Perspective." *Environmental Health Perspectives* 110 2 (2002): 259-64. Print.

Barber, Katrine. *Death of Celilo Falls*. The Emil and Kathleen Sick lecture-book series in western history and biography. Seattle: Center for the Study of the Pacific Northwest in association with University of Washington Press, 2005. Print.

Borgmann, Albert. *Technology and the Character of Contemporary Life: A Philosophical Inquiry*. Chicago: University of Chicago Press, 1984. Print.

Bullard, Robert D. *Confronting Environmental Racism: Voices from the Grassroots*. 1st ed. Boston, MA: South End Press, 1993. Print.

Cohen, Goel. *Technology Transfer: Strategic Management in Developing Countries*. New Delhi India; Thousand Oaks CA: Sage Publications, 2004. Print.

Dietz, Henry A. *Poverty and Problem-Solving under Military Rule: The Urban Poor in Lima, Peru*. Latin American monographs no. 51. Austin: University of Texas Press, 1980. Print.

Frye, Marilyn. *The Politics of Reality: Essays in Feminist Theory*. Crossing Press feminist series. Trumansburg, NY: Crossing Press, 1983. Print.

Grijalva, James M. *Closing the Circle: Environmental Justice in Indian Country*. Durham, NC: Carolina Academic Press, 2008. Print.

Grinde, Donald A., and Bruce E. Johansen. *Ecocide of Native America: Environmental Destruction of Indian Lands and Peoples*, 1st ed. Sante Fe, NM: Clear Light, 1995. Print.

Harkin, Michael Eugene, and David Rich Lewis. *Native Americans and the Environment: Perspectives on the Ecological Indian*. Lincoln: University of Nebraska Press, 2007. Print.

Isleta, Pueblo of. "Surface Water Quality Standards." 2002. Print.

Krech, Shepard. *The Ecological Indian: Myth and History*. 1st ed. New York: W.W. Norton & Company, 1999. Print.

LaDuke, Winona. *All Our Relations: Native Struggles for Land and Life*. Cambridge, MA Minneapolis, MN: South End Press; Honor the Earth, 1999. Print.

Mutz, Kathryn M., Gary C. Bryner, and Douglas S. Kenney. "Justice and Natural Resources: Concepts, Strategies, and Applications," Washington, DC, 2002. xxxvii, 368 p. Island Press.

NEJAC. "Ensuring Risk Reduction in Communities with Multiple Stressors: Environmental Justice and Cumulative Risks/Impacts." Washington, DC: National Environmental Justice Advisory Council, 2004. Print.

Ranco, Darren, and Ann Fleder. "Tribal Environmental Sovereignty: Cultural Appropriate Protection or Paternalism?" *Journal of Natural Resources and Environmental Law* 19 1 (2005): 35-58. Print.

Royster, Judith. "Environmental Federalism and the Third Sovereign: Limits on State Authority to Regbulate Water Quality in Indian Country." *Water Resources Update* 105 (1996). Print.

---. "A Primer on Indian Water Rights: More Questions than Answers." *Tulsa Law Journal* 30 (1994-1995): 61-104. Print.

Sanders, Marren. "Clean Water in Indian Country: The Risks (and Rewards) of Being Treated in the Same Manner as a State." *William Mitchell Law Review* Forthcoming (2009). Print.

Selinger, Evan. "Does Microcredit Empower? Reflections on the Grameen Bank Debate." *Human Studies* 31 (2008): 27-41. Print.

Shiva, Vandana. "The Violence of the Green Revolution: Third World Agriculture, Ecology, and politics". London ; Atlantic Highlands, N.J., USA, 1991. 264 p. Zed Books.

Shrader-Frechette, K. S. *Environmental Justice: Creating Equality, Reclaiming Democracy.* Environmental Ethics and Science Policy Series. Oxford ; New York: Oxford University Press, 2002. Print.

Singer, Michelle. "Inuit Life Threatened By Climate Change," 2007. CBS Interactive Inc. 2009.

Sturgeon, Noël. *Environmentalism in Popular Culture: Gender, Race, Sexuality, and the Politics of the Natural.* Tucson: University of Arizona Press, 2009. Print.

Sullivan, Nicholas P. *You Can Hear Me Now: How Microloans and Cell Phones are Connecting the World's Poor to the Global Economy*, 1st ed. San Francisco, CA: Jossey-Bass, 2007. Print.

Tarbell, Alice, and Mary Arquette. "Akwesasne: A Native Community's Resistance to Cultural and Environmental Damage." *Reclaiming the Environmental Debate*. Ed. Hofrichter, Richard. Cambridge, MA: MIT Press, 2001. 93-113. Print.

Thorpe, Grace "Our Homes Are Not Dumps." *Defending Our Mother: Native American Perspectives on Environmental Justice*. Ed. Weaver, Jace. Maryknoll, NY Orbis Books, 1996. 47-58. Print.

Tiles, Mary, and Hans Oberdiek. *Living in a Technological Culture: Human Tools and Human Values.* Philosophical Issues in Science. London; New York: Routledge, 1995. Print.

Walker, J. Samuel, and U.S. Nuclear Regulatory Commission. *The Road to Yucca Mountain: The Development of Radioactive Waste Policy in the United States*. Berkeley: University of California Press, 2009. Print.

Weaver, Jace. *Defending Mother Earth: Native American Perspectives on Environmental Justice*. Maryknoll, NY: Orbis Books, 1996. Print.

---. "Notes from a Miner's Canary." *Defending Mother Earth: Native American Perspectives on Environmental Justice*. Ed. Weaver, Jace. Maryknoll, NY: Orbis Books, 1996. 1-28. Print.

Westra, Laura, and Bill E. Lawson. *Faces of Environmental Racism: Confronting Issues of Global Justice*. Studies in Social, Political, and Legal Philosophy, 2nd ed. Lanham, Md.: Rowman & Littlefield Publishers, 2001. Print.

A Recession in the Economy of Trust[1]

Thomas A. Easton[2]

Atechnology is now moving from industry to the home with the potential to change the way we live at a very fundamental level. It may even reduce the level of trust we extend each other and the degree to which we trust the evidence of our senses.

The last time a technology with such drastic effects appeared was in the 1970s, when computers were huge, expensive, and used only by large organizations. But that was when the first primitive personal computers (PCs) appeared. At that time the PC was of interest mainly to hobbyists. But within twenty years, people were beginning to work, play, shop, socialize, and even steal online. Today e-commerce is a major sector of the economy, telework is a serious alternative for millions of workers, online gaming is a major industry, we worry about scamming, phishing, and identity theft, and some people debate the validity of online or virtual relationships.

The idea that technology can lead to rapid and widespread change is not new. Nor is it new that some people object to that change, fearing that it outruns the ability of people and society to adjust to it. There are grounds for such objections, of course. Critics of technological progress often point to military technology to support their claim that technological progress outstrips social progress, and indeed it is now possible to kill huge numbers of people at one blow. On the other hand, technological progress in communications and transportation has meant a huge increase in knowledge about other parts of the world. This knowledge includes knowledge about humanitarian disasters, as after the 2004 tsunami in Southeast Asia, and the technology enables an enormous increase in fund-raising and aid delivery. Even a dedicated critic of technology would have to say that this is an example of technological progress and social progress going together.

It was noted years ago that technology has both good and bad effects, and they cannot be separated.[3] The contrast between good and bad is greatest for those new technologies called "disruptive technologies." Such technologies change the rules. They destroy businesses. They make industries and jobs obsolete. They can even cut the funds available to governments to support schools and maintain roads or—as in the case of the automobile—force government to find the funds to pay for new services (such as roads). At the same time, however, there are more positive effects. As businesses, industries, and jobs go away, new ones appear, and historically the new ones more than make up for the old ones that have vanished. It may take time, and the transition period may be painful, but it happens. In the long run, employment has steadily increased. So has society's wealth and human welfare.

3D Printing

The newest disruptive technology is 3D printing, also known as rapid prototyping and instant manufacturing.[4] It is based on the idea that any three-dimensional object can be viewed as a stack of two-dimensional slices. A 3D printer (also known as a "fabber," short for fabricator) prints the slices, one on top of another, until it produces (or "fabs") the whole object. For "ink" or raw material, it uses powders (similar to copier toner) or liquid plastics or even pastes. It has a technique to solidify the ink as each layer is laid down. If there is internal structure—hollow spaces, pieces of different colors or consistency—that is all faithfully reproduced.

The technology has been in use in industry for nearly twenty years. Z-Corporation's machines use the powder method; the powder layers are solidified when a form of ink-jet printer with four print heads deposits a thin layer of binder material (in three colors and clear). The printed items must be treated with a liquid "infiltrant" that hardens the material enough to support handling. The items are not, however, sturdy enough to install in products intended for sale and use. They are prototypes, demonstration models, and the originals around which molds can be formed for making plastic, metal, and ceramic items for actual use. These uses are valuable enough to make the company a growing concern.

Stratasys machines use a different method—Fused Deposition Modeling (FDM). In a heated chamber, two moving heads extrude thin streams of melted plastic (in several colors) to form each layer of an object being printed. The use of melted plastic gives the items much greater solidity.

What do these machines cost? The Zprinter 450 was touted as a price breakthrough when it came out in 2007—and it sells for a bit less than $40,000. The smaller Zprinter 310, which uses only one color of binder, costs about $25,000. Stratasys's FDM200mc sells in the same range. Stratasys's Dimension line starts at under $19,000 and goes up to $33,000. Such prices don't sound like much of a breakthrough until we note that high-end rapid-prototyping machines can cost as much as $500,000.

Even the low end of this price range is a bit much for most home and small business budgets, and that's a shame. 3D printing has an enormous amount of appeal to anyone who likes to make things. Museums use them to make replicas of bones and fossils. Artists use them to make sculptures. Medical labs have used versions of the technology to print precisely shaped bone implants from bone-like material and hearing aids that precisely fit a patient's ear canal. Under development are rapid prototyping machines for dental labs and even dentists' offices, where they could make crowns and false teeth. A machine based on a modified inkjet printer has produced sheets of biodegradable gel with embedded cells; the eventual aim is to make on demand custom-designed tissues and organs for use in transplants. Researchers have even begun to develop techniques for "printing" skin and blood vessels.[5] If everything pans out, the future of health care should be very interesting.

Considering the high prices of much medical equipment, the high prices of industrial 3D printers are no obstacle for health care. For the rest of us, these machines are out of reach. However, smaller machines with much smaller price tags are now available, and they will change the nature of the game. Once the technology has matured, you will be able to make just about anything that fits inside your printer. It sounds like science fiction—in fact, it sounds a lot like a *Star Trek* replicator!

One engineering professor is working on one large enough to print out houses (using concrete as "ink") at much less than normal construction cost.[6] Given that, what can we say about the future? 3D printers will eventually come in many sizes, use many raw materials, and be able to make a huge variety of items. The consumer will be happy. But new technologies always deliver both plusses and minuses. In this case the minuses appear as soon as we consider that if we are making things at home, those who previously made them in factories for sale in stores will have little left to do. Businesses and even industries will be forced to shut down or drastically change their way of doing business. Countries whose economies depend on exports of printable items will experience recessions or even depressions. Workers will lose jobs, or be obliged to learn new jobs. And national, state, and municipal governments will not see the funds they are accustomed to receiving from sales taxes (including VAT and GST taxes). These funds are used to provide vital public services (such as schools, fire departments, and police) and will have to be replaced somehow.

Taking 3D-Printing Home

When rapid prototyping or 3D printing reaches the home market, it will give people the power to make a great many things they now have to pay for. Even more intriguingly, because it prints *things*, not just pictures, it can print its own parts. The first 3D printer can then become two, which quickly become four, and so on. Costs can become extraordinarily low, and the means of production can spread quickly throughout society. Since some users will tinker with their machines to improve them, the best improvements will spread fastest, in a process akin to Darwinian natural selection.

This is the basic idea behind the RepRap project,[7] whose founder, Adrian Bowyer of the University of Bath, says "the replicating rapid prototyping machine will allow the revolutionary ownership, by the proletariat, of the means of production. But it will do so without all that messy and dangerous revolution stuff, and even without all that messy and dangerous industrial stuff. Therefore I have decided to call this process Darwinian Marxism."

The ultimate goal of the project is a von Neumann machine, a machine that can reproduce itself. The prospect of such machines makes some people nervous, for the one thing that seems likely to keep robots from ever taking over the world is their dependence on humans to make them. At the moment, however, the von Neumann machine goal is a long way off. RepRap machines are made of many plastic parts the printer can produce, using the FDM technique, but many other parts must be supplied by a human being. These include metal rods, screws, motors, power supplies, and computer chips.

The RepRap project makes available to all who would like to build their own 3D printer parts lists (the parts should cost less than $600) and instructions for building and programming the machine. And there's no charge. The project's motto is "wealth without money," and it begins with the project itself. For future users, the motto means they will be able to satisfy a great many needs and desires without worrying about whether they can afford to pay for them. Eventually, it may mean that human civilization is built upon an industrial infrastructure that continually builds, rebuilds, and improves itself, without the need for investors.

Unfortunately, the RepRap machines are not yet ready to take home. For that we must look to two other 3D printer projects, Fab@Home and Desktop Factory.

In 2006, Hod Lipson of Cornell University launched the Fab@Home project with PhD student Evan Malone. By January 2007, they were able to announce that their "Freeform" fabber—about the size of a microwave oven—could be assembled for about $2,400. The parts list, with a list of suppliers, instructions on how to build and operate it, and all the necessary software are available—for free—from their website.[8] As with the RepRap machine, the idea is that people should feel free to modify and—hopefully—improve the machine. Also like the RepRap machine, it uses a version of Fused Deposition Modelling, but in addition to plastic, it can print using PlayDoh, cheese, silicone caulk, plaster, chocolate, cake frosting, metal-impregnated plastics (for printing wires), and other soft substances that will harden quickly.

By the summer of 2007, it was already possible to buy fully assembled versions of the Fab@Home fabber for about $3,600. This is a great deal cheaper than any of the industrial machines, and though performance is not as good as with the industrial machines (the Fab@Home fabber is slower, and the surfaces of the objects it makes are not as smooth), with its ability to use many different materials, it has an astonishing versatility. It will only improve over time. Eventually… Well, Evan Malone has already built a version of the machine that uses a rack of syringes and can make things out of several materials at the same time. He has used it to make a working battery, and his ultimate goal is to use his fabber to make a complete, working robot. If it can do that, of course, it can probably make another fabber, just as the RepRap folks intend.

Idealab was started in 1996 "to create and operate pioneering technology companies." In 2004, it gave birth to Desktop Factory,[9] whose goal "is to one day make 3D printing as common in offices, factories, schools and homes as laser printers are today. Just as desktop publishing exploded as prices dropped and laser printers became ubiquitous, so too will new uses for 3D printing emerge as devices become inexpensive and widely available."

The Desktop Factory 125ci 3D printer easily fits on a desktop and weighs less than ninety pounds. It can make things up to five inches on a side with layers a hundredth of an inch thick, slightly thicker than those laid down by the industrial machines. Speed of printing is comparable to that of the industrial machines. For raw material, it uses a proprietary plastic powder that can be fused by light from a relatively inexpensive halogen bulb to make things sturdy "enough to throw across a conference table."

The Wave of the Future?

Many people are already enchanted by the prospect of being able to have their own 3D printer,[10] but the price has to come down before large numbers of people buy into the technology. Another obstacle is that present printers such as RepRap and Fab@Home require a degree of comfort with do-it-yourself geekery that most people just don't have. But technology evolves. Prices drop. Gadgets become easier to use. And these things can happen rapidly. The PC went from something that had to be programmed by the user before it could do anything useful to an economic revolution in just two decades. 3D printing may go at least that far in less time.

It's worth noting that the Fab@Home and Desktop Factory fabbers aren't really that expensive. If we look back at the 1970s, at the threshold of the PC revolution, home computer prices look cheaper, but that doesn't allow for inflation. In 1972, the HP9830, the "first desktop all-in-one," sold for $5,975 . In 2008 dollars, that's $30,000, in the same range as the cheaper industrial-grade 3D printers. In 1975, the Altair sold (assembled) for $621, which works out to about $2,400. In 1977, the Apple II sold for $1,295, or $4,400 in 2008 dollars. They compare very well to Fab@Home's $3,600 (assembled) and Desktop

Factory's $5,000. If 3D printers follow a price history anything like the PC's, they will be become much cheaper and enormously more capable over the next few years. That is when they will begin to have serious social effects.

The Problem of Authentication

One of the things the PC has done is to force us to question the authenticity of many cultural artifacts. This is of particular note in connection with photos and sound and video recordings. In the pre-computer past, it was possible but difficult to alter photos, or to edit a sound tape by splicing out portions, or to dub a tape with fragments from others to make someone seem to say something they never did. But such manipulations were relatively easy to spot. Today, photos and sound files can be edited almost as easily as can text files, and the necessary software is widely available. With photos, we actually say that an edited picture has been "Photoshopped." We don't have an equivalent word for sound files, but very effective free software is available. The signs of editing remain detectable, but they are less obvious, especially to the untrained eye or ear.

So far, video is harder to manipulate, but video-editing software exists and people are beginning to realize that just as they cannot trust a photo or sound recording to represent faithfully objective reality, even video may be suspect. Since doctored images and recordings have shown up on national news programs, people have already begun to trust the news less than they used to.

And trust is important to the smooth working of society. In fact, it is so important that people actually speak of an "economy of trust"[11] that facilitates political, financial, and other interactions. We cannot vote for a politician we cannot trust to look out for our interests. We cannot put our money in a bank or a mutual fund that we cannot trust to keep it safe and even growing for us. We cannot buy insurance from a company we cannot trust to honor its promises. Politicians, banks, and insurance companies (and some other businesses) therefore try very hard to convince us that they are trustworthy. Think of Allstate's "You're in Good Hands" slogan! They have an easier time of it because we are brought up to trust our parents, teachers, ministers and other adults, and despite political and business scandals, that training sticks. We are willing to trust, and even if we turn cynical toward politicians and businesses, we remain willing to trust our friends, relatives, and coworkers.

We are also, of course, willing to trust the evidence of our senses. There was a time when that meant trusting that what we saw or heard for ourselves was real and undoctored. Today much of what we see and hear is electronically mediated. We see pictures in magazines, on television, and on websites. We hear recordings of what people said. We see video clips. And those of us who are aware of how easy it is to manipulate electronic images and recordings are less willing to trust them. Sometimes the lesson is a hard one.

The lesson is likely to get harder. The 3D printer will extend our need to doubt authenticity to physical objects, not just digital files. Unscrupulous dealers in antiques and collectibles will be scanning valuable items and printing out duplicates that look and feel authentic. Virginia artist Cortney Skinner[12] suggests that art thieves might create duplicates of items in museums or galleries. Later they might steal the original and leave the duplicate in its place, so that no one is aware the theft even occurred.[13] Of course, a thief could also make a copy and offer it as the stolen original. If the copy is good enough, it should pass.

3D printers are already being used to make duplicates of fossils for research, display, and educational purposes. Among the varieties of scientific fraud is the manufacturing of experimental evidence, which may before long come to include the making of entirely fictional "fossils." This could also be done by nonscientists who wish to prove a point such as the creationist claim that humans and dinosaurs walked the Earth at the same time. I can imagine fake fossil footprints, human beside dinosaur; a fossil dinosaur with human finger bones in its tummy (yes, dinosaur fossils have been found with recent meals still in place); a dinosaur bone with an embedded stone spear point; or a dinosaur coprolite (fossilized dung) with embedded bits of human bone.

I can also imagine fake evidence of alien visitations. Years ago, the *National Enquirer* asked me to give my opinion of a photo of what they said was a sprig of alien vegetation from a clot of mud on a flying saucer's landing gear. It certainly looked weird enough, but I had to tell them it was only a scrap of Earthly fungus, complete with a scale bar that said it was about a tenth of a millimeter long. A 3D printer could produce a much more convincing bit of alien vegetation, garbage, equipment, or even bones. It could also produce a fossil fairy, complete with wings, to satisfy those who crave proof of the existence of elves and their kin. Or perhaps a dragon or mermaid! The *Weekly World News*[14] was infamous for its Photoshopped "evidence" of such creatures. Think of what they could have done with a 3D printer!

It will for some time be impossible to match the materials of original and fake in both substance and texture, so such fakes will be fairly easy to spot. But often what goes on display in a museum is no more than a cast of the original fossil, and textbooks show only photos. A fake fossil or bit of alien debris could pass for awhile, especially among those who already believe in the cause.

A Crisis of Trust

Trust even in the evidence of our senses is no longer something we can take for granted. It never was, of course, but it was not so long ago that our distrust took the form of "Did he mean what he said?" rather than "Did he really say (or do) that?" Today, evidence such as photos and sound or video recordings can be manufactured to "prove" anything we like. The technology to edit images and recordings is more and more available and people are more and more aware of what it can do.

3D printing will very quickly add physical objects to the list of items whose authenticity can no longer be taken for granted. Is a fossil real or has it been faked up to support a bogus belief? Is a spare part genuine or is it a terrorist's booby-trap waiting to go off? Are products licensed, or are they counterfeit? Are artworks and other collectibles genuine? Fakes exist already, but with high-quality 3D printing widely available—as it seems very likely to be within a few years—there will be a great many more of them.

It might seem ingenuous that we still take authenticity for granted despite the ease of copying or modifying digital files and despite the current availability of counterfeit fashion, watches, shoes, electronics, and other products. But if we do not—if we refuse to trust the people with whom we deal in daily life—we must feel obliged to check the authenticity of every item we buy and every photo or video we see or tape we hear. In principle it would be possible to do this, especially because Photoshopping (and its video and audio equivalents) leaves detectable signs, and digital watermarking can authenticate files. Counterfeiting of physical products is already widespread, and countermeasures such as micro-taggants[15] are available. But we don't check, except in certain cases. Cashiers use a special marking pen on large-denomination bills. Ebay buyers check a seller's reputation rating. And young people Google prospective dates.

Trust plays an essential part in our social and commercial interactions. In fact, we exchange trust in much the same way we exchange goods for money. The "economy of trust" is thus akin to the "economy of money" that is usually meant by the word "economy." It works because even though trust is often enough betrayed (checks bounce, too), it is more often *not* betrayed.

The economy of trust has not yet been hampered by digital fakery. We remain willing to trust politicians to look after our interests (though we are cynical about their promises), insurance companies to pay claims, and banks to keep money safe and pay interest. And when we are shown an object, or even a picture of the object, we believe the fossil, sculpture, or product is real, not faked.

If 3D printing develops in such a way that we can no longer believe in the authenticity of physical objects, the blow to trust will be much more profound than past revelations of faked photos, voice recordings, and videos. Trust itself will be devalued to the point where it is not just metaphor to speak of a recession in the economy of trust.

On the other hand, it is possible—at least for now—to tell fake from real. If we do not wish to give up our ability to trust our fellows, if we wish to prevent that recession from happening, we must put more effort into developing ways (such as microtaggants) to authenticate things.

Notes

1. Adapted from *The 3D Printing Revolution: Social and Economic Impacts* (in prep.).
2. Professor of Science, Thomas College, Waterville, ME. Contact: Easton@thomas.edu.
3. E. G. Mesthene, "The Role of Technology in Society," *Technology and Culture*, vol. 10, No. 4, pp. 489-536, 1969.
4. I. Amato, "Instant Manufacturing," *Technology Review*, Vol. 106, No. 9, pp. 56-62, November 2003.
5. K. Green, "Printing Blood Vessels," *Technology Review,* January 20, 2006 (http://www.technologyreview.com/BioTech/wtr_16168,306,p1.html)
6. See http://www.contourcrafting.org.
7. See http://reprap.org/bin/view/Main/ShowCase.
8. See http://fabathome.org.
9. See http://www.desktopfactory.com.
10. Chris Morrison, "3D Printing for the Rest of Us," *Business 2.0*, vol. 8, No. 8, pp. 46-47, September 2007.
11. K. Arrow, "The Economy of Trust," *Religion and Liberty*, vol 16, No. 3, pp. 3, 12, Summer 2006.
12. Personal communication.
13. The perfect defense against this, of course, is for the museum itself to make copies and only put those on display. Then the joke could be on the thieves!
14. The print edition is defunct, but the tabloid lives on at http://weeklyworldnews.com/.
15. See e.g. http://www.microtaggant.com/whataretaggants.htm.

Contributors

John Barker, Assistant Professor of Philosophy at University of Illinois Springfield.

William Cornwell, Associate Professor of Philosophy at Salem State College.

Thomas A. Easton, Professor of Science at Thomas College.

Grant Havers, Associate Professor of Philosophy and Political Studies at Trinity Western University.

Alan Kim, Visiting Assistant Professor of Philosophy at Colgate University.

Christopher Vasillopulos, Professor of Political Science at Eastern Connecticut State University.

Kyle Powys Whyte, Visiting Assistant Professor of Philosophy at Michigan State University.

Thomas R. Winpenny, Professor of History at Elizabethtown College.

Fani Zlatarova, Professor of Computer Sciences at Elizabethtown College.

Gabriel R. Ricci is professor of humanities and the chair of the Department of History at Elizabethtown College. He is the author of *Time Consciousness: The Philosophical Uses of History* and the editor of Transaction's much-admired *Religion and Public Life* series.